Electromagnetism

Problems and solutions

Electromagnetism

Problems and solutions

Carolina C Ilie
State University of New York at Oswego, USA

Zachariah S Schrecengost
State University of New York at Oswego, USA

Morgan & Claypool Publishers

Rights & Permissions
To obtain permission to re-use copyrighted material from Morgan & Claypool Publishers, please contact info@morganclaypool.com.

ISBN   978-1-6817-4429-2 (ebook)
ISBN   978-1-6817-4428-5 (print)
ISBN   978-1-6817-4431-5 (mobi)

DOI   10.1088/978-1-6817-4429-2

Version: 20161101

IOP Concise Physics
ISSN 2053-2571 (online)
ISSN 2054-7307 (print)

A Morgan & Claypool publication as part of IOP Concise Physics
Published by Morgan & Claypool Publishers, 40 Oak Drive, San Rafael, CA, 94903 USA

IOP Publishing, Temple Circus, Temple Way, Bristol BS1 6HG, UK

To my family, my mentors, and my students — CCI

To my friends, family, and mentors — ZSS

Contents

6   Magnetic fields in matter   6-1

Preface

We wrote this book of problems and solutions having in mind the undergraduate student—sophomore, junior, or senior—who may want to work on more problems and receive immediate feedback while studying. The authors strongly recommend the textbook by David J Griffiths, *Introduction to Electrodynamics*, as a first source manual, since it is recognized as one of the best books on electrodynamics at the undergraduate level. We consider this book of problems and solutions a companion volume for the student who would like to work on more electrostatic problems by herself/himself in order to deepen their understanding and problems solving skills. We add brief theoretical notes and formulae; for a complete theoretical approach we suggest Griffiths' book. Every chapter is organized as follows: brief theoretical notes followed by the problem text with the solution. Each chapter ends with a brief bibliography.

We plan to write a second volume on electrodynamics, which will start with Maxwell's equations and the conservation laws, and then discuss electromagnetic (EM) waves, potentials and fields, radiation, and relativistic electrodynamics.

We follow here the notation of Griffiths, and use $\vec{\imath}$ for the vector from a source point $\vec{r}\,'$ to the field point $\vec{r}$. Please note that $\hat{\imath} = \dfrac{\vec{\imath}}{\imath} = \dfrac{\vec{r} - \vec{r}\,'}{|\vec{r} - \vec{r}\,'|}$ and, as you see, this notation already greatly simplifies complex equations, but you need to be careful with your notation, in particular if you only use cursive or typed letters. Also, we use the same notation s for the distance to the z-axis in cylindrical coordinates as is used in Griffiths' book.

The chosen units are SI units—the international system. The reader should be aware that other books may employ either the Gaussian system (CGS) or the Heaviside–Lorentz (HL) system. The Coulomb force in each of the systems is as follows,

SI system:

$$\vec{F} = \frac{1}{4\pi\varepsilon_0} \frac{q_1 q_2}{\imath^2} \hat{\imath}$$

CGS:

$$\vec{F} = \frac{q_1 q_2}{\imath^2} \hat{\imath}$$

HL:

$$\vec{F} = \frac{1}{4\pi} \frac{q_1 q_2}{\imath^2} \hat{\imath}$$

Some of the problems are typical practice problems with the pedagogical role of improving understanding and problem solving skills. Several of the problems presented here appear in a variety of undergraduate textbooks on EM as they are classic examples; however, we felt it would be incomplete to omit these problems as they are fundamental to the study of EM. We also present problems that are more general in nature, which may be a bit more challenging. We tried to maintain a balance between the two types of problems, and we hope that the readers will enjoy this variation and have as much thrill and excitement as we had while creating and solving these problems.

Acknowledgements

We want to thank to Dr Ilie's students, Nicholas Jira, Vincent DeBiase, Ian Evans, and Andres Inga, who contributed to the editing (typing) of this book. We are particularly grateful to our illustrator, Julia D'Rozario, for making all of the figures. We thank Dr Ildar Sabirianov for providing useful suggestions. We thank the administration at SUNY Oswego and the office of Research and Individualized Student Experiences for overall support. We are grateful to Dr Peter Dowben, from the University of Nebraska at Lincoln, who thought that such a project has a niche. A thought of appreciation to Dr Charles Ebner, from the Ohio State University for his perfect Electrodynamics course. Also many thanks to our editors, Joel Claypool, Publisher at Morgan & Claypool Publishers, Jeanine Burke, Consulting Editor at the IOP Concise Physics e-book program, and Jacky Mucklow, Production Team Manager at the Institute of Physics. Lastly, we thank to our families and friends for their sense of humor, encouragement, and for keeping us sane and happy.

About the authors

Carolina C Ilie

Carolina C Ilie is an Associate Professor with tenure at the State University of New York at Oswego. She taught Electromagnetic Theory for almost ten years and designed various problems for her students' exams, group work, and quizzes. Dr Ilie obtained her PhD in Physics and Astronomy at the University of Nebraska at Lincoln, an MSc in Physics at Ohio State University and another MSc in Physics at the University of Bucharest, Romania. She received the President's Award for Teaching Excellence in 2016 and the Provost Award for Mentoring in Scholarly and Creative Activity in 2013. She lives in Central New York with her spouse, also a physicist, and their two sons.

Photograph courtesy of James Russell/SUNY Oswego Office of Communications and Marketing.

Zachariah S Schrecengost

Zachariah S Schrecengost is a State University of New York alumnus. He graduated summa cum laude with a BS degree having completed majors in Physics, Software Engineering, and Applied Mathematics. He took the Advanced Electromagnetic Theory course with Dr Ilie and was thrilled to be involved in creating this book. He brings to the project both the fresh perspective of the student taking electrodynamics, as well as the enthusiasm and talent of an alumnus who is an electrodynamics and upper level mathematics aficionado. Mr Schrecengost works as a software engineer in Syracuse and is preparing to begin his graduate school studies in physics.

Julia R D'Rozario

Julia R D'Rozario (*illustrator*) graduated from the State University of New York at Oswego in December 2016 where she completed a BS in Physics and a BA in Cinema and Screen Studies, and completed a minor in Astronomy by May 2016. She completed the Advanced Electromagnetic Theory course with Dr Ilie and has much experience of the arts through her career in film. Ms D'Rozario contributes her knowledge of electrodynamics and her talent in drawing using Inkscape software. Her future aim is to attend graduate school and continue to combine her passions for physics and cinema.

Electromagnetism
Problems and solutions
Carolina C Ilie and Zachariah S Schrecengost

Chapter 1

Mathematical techniques

There are a variety of mathematical techniques required to solve problems in electromagnetism. The aim of this chapter is to provide problems that will build confidence in these techniques. Concepts from vector calculus and curvilinear coordinate systems are the primary focus.

1.1 Theory

1.1.1 Dot and cross product

Given vectors $\vec{A} = A_x\hat{x} + A_y\hat{y} + A_z\hat{z}$ and $\vec{B} = B_x\hat{x} + B_y\hat{y} + B_z\hat{z}$

$$\vec{A} \cdot \vec{B} = A_xB_x + A_yB_y + A_zB_z = AB\cos\theta$$

$$\vec{A} \times \vec{B} = \begin{vmatrix} \hat{x} & \hat{y} & \hat{z} \\ A_x & A_y & A_z \\ B_x & B_y & B_z \end{vmatrix} \quad \text{with} \quad \left|\vec{A} \times \vec{B}\right| = AB\sin\theta$$

where $A = |\vec{A}| = \sqrt{A_x^2 + A_y^2 + A_z^2}$, $B = |\vec{B}| = \sqrt{B_x^2 + B_y^2 + B_z^2}$, and θ is the angle between $\vec{A}$ and $\vec{B}$.

1.1.2 Separation vector

This notation is outlined by David J Griffiths in his book *Introduction to Electrodynamics* (1999, 2013). Given a source point $\vec{r}'$ and field point $\vec{r}$, the separation vector points from $\vec{r}'$ to $\vec{r}$ and is given by

$$\vec{r} = \vec{r} - \vec{r}' = (x - x')\hat{x} + (y - y')\hat{y} + (z - z')\hat{z}$$

doi:10.1088/978-1-6817-4429-2ch1

and the unit vector pointing from $\vec{r}'$ to $\vec{r}$ is

$$\hat{r} = \frac{\vec{r}}{r} = \frac{\vec{r} - \vec{r}'}{|\vec{r} - \vec{r}'|} = \frac{(x - x')\hat{x} + (y - y')\hat{y} + (z - z')\hat{z}}{\sqrt{(x - x')^2 + (y - y')^2 + (z - z')^2}}.$$

As explained by Griffiths, this notation greatly simplifies later equations.

1.1.3 Transformation matrix

Given vector $\vec{A} = A_x\hat{x} + A_y\hat{y} + A_z\hat{z}$ in coordinate system K, the components of $\vec{A}$ in coordinate system K' are determined by rotational matrix R given by

$$R = \begin{pmatrix} R_{xx} & R_{xy} & R_{xz} \\ R_{yx} & R_{yy} & R_{yz} \\ R_{zx} & R_{zy} & R_{zz} \end{pmatrix}$$

with

$$\begin{pmatrix} A'_x \\ A'_y \\ A'_z \end{pmatrix} = R \begin{pmatrix} A_x \\ A_y \\ A_z \end{pmatrix}.$$

1.1.4 Gradient

Given a scalar function T, the gradients for various coordinate systems are given below.

Cartesian

$$\nabla T = \frac{\partial T}{\partial x}\hat{x} + \frac{\partial T}{\partial y}\hat{y} + \frac{\partial T}{\partial z}\hat{z}$$

Cylindrical

$$\nabla T = \frac{\partial T}{\partial s}\hat{s} + \frac{1}{s}\frac{\partial T}{\partial \phi}\hat{\phi} + \frac{\partial T}{\partial z}\hat{z}$$

Spherical

$$\nabla T = \frac{\partial T}{\partial r}\hat{r} + \frac{1}{r}\frac{\partial T}{\partial \theta}\hat{\theta} + \frac{1}{r \sin \theta}\frac{\partial T}{\partial \phi}\hat{\phi}$$

1.1.5 Divergence

Given vector function $\vec{v}$, the divergences for various coordinate systems are given below.

Cartesian

$$\nabla \cdot \vec{v} = \frac{\partial v_x}{\partial x} + \frac{\partial v_y}{\partial y} + \frac{\partial v_z}{\partial z}$$

Cylindrical

$$\nabla \cdot \vec{v} = \frac{1}{s}\frac{\partial}{\partial s}(sv_s) + \frac{1}{s}\frac{\partial v_\phi}{\partial \phi} + \frac{\partial v_z}{\partial z}$$

Spherical

$$\nabla \cdot \vec{v} = \frac{1}{r^2}\frac{\partial}{\partial r}\left(r^2 v_r\right) + \frac{1}{r\sin\theta}\frac{\partial}{\partial \theta}\left(\sin\theta\, v_\theta\right) + \frac{1}{r\sin\theta}\frac{\partial v_\phi}{\partial \phi}$$

1.1.6 Curl

Given vector function $\vec{v}$, the curls for various coordinate systems are given below.

Cartesian

$$\nabla \times \vec{v} = \left(\frac{\partial v_z}{\partial y} - \frac{\partial v_y}{\partial z}\right)\hat{x} + \left(\frac{\partial v_x}{\partial z} - \frac{\partial v_z}{\partial x}\right)\hat{y} + \left(\frac{\partial v_y}{\partial x} - \frac{\partial v_x}{\partial y}\right)\hat{z}$$

Cylindrical

$$\nabla \times \vec{v} = \left(\frac{1}{s}\frac{\partial v_z}{\partial \phi} - \frac{\partial v_\phi}{\partial z}\right)\hat{s} + \left(\frac{\partial v_s}{\partial z} - \frac{\partial v_z}{\partial s}\right)\hat{\phi} + \frac{1}{s}\left[\frac{\partial}{\partial s}(sv_\phi) - \frac{\partial v_s}{\partial \phi}\right]\hat{z}$$

Spherical

$$\nabla \times \vec{v} = \frac{1}{r\sin\theta}\left[\frac{\partial}{\partial \theta}(\sin\theta\, v_\phi) - \frac{\partial v_\theta}{\partial \phi}\right]\hat{r} + \frac{1}{r}\left[\frac{1}{\sin\theta}\frac{\partial v_r}{\partial \phi} - \frac{\partial}{\partial r}(rv_\phi)\right]\hat{\theta}$$
$$+ \frac{1}{r}\left[\frac{\partial}{\partial r}(rv_\theta) - \frac{\partial v_r}{\partial \theta}\right]\hat{\phi}$$

1.1.7 Laplacian

Given a scalar function T, the Laplacians for various coordinate systems are given below.

Cartesian

$$\nabla^2 T = \frac{\partial^2 T}{\partial x^2} + \frac{\partial^2 T}{\partial y^2} + \frac{\partial^2 T}{\partial z^2}$$

Cylindrical

$$\nabla^2 T = \frac{1}{s}\frac{\partial}{\partial s}\left(s\frac{\partial T}{\partial s}\right) + \frac{1}{s^2}\frac{\partial^2 T}{\partial \phi^2} + \frac{\partial^2 T}{\partial z^2}$$

Spherical

$$\nabla^2 T = \frac{1}{r^2}\frac{\partial}{\partial r}\left(r^2\frac{\partial T}{\partial r}\right) + \frac{1}{r^2\sin\theta}\frac{\partial}{\partial \theta}\left(\sin\theta\frac{\partial T}{\partial \theta}\right) + \frac{1}{r^2\sin^2\theta}\frac{\partial^2 T}{\partial \phi^2}$$

1.1.8 Line integral

Given vector function $\vec{v}$ and path $\mathcal{P}$, a line integral is given by

$$\int_{\vec{a}\,\mathcal{P}}^{\vec{b}} \vec{v} \cdot d\vec{\ell},$$

where $\vec{a}$ and $\vec{b}$ are the end points, and $d\vec{\ell}$ is the infinitesimal displacement vector along $\mathcal{P}$. In Cartesian coordinates $d\vec{\ell} = dx\,\hat{x} + dy\,\hat{y} + dz\,\hat{z}$.

1.1.9 Surface integral

Given vector function $\vec{v}$ and surface $\mathcal{S}$, a surface integral is given by

$$\int_{\mathcal{S}} \vec{v} \cdot d\vec{a},$$

where $d\vec{a}$ is the infinitesimal area vector that has direction normal to the surface. Note that $d\vec{a}$ always depends on the surface involved.

1.1.10 Volume integral

Given scalar function T and volume $\mathcal{V}$, a volume integral is given by

$$\int_{\mathcal{V}} T\,d\tau,$$

where $d\tau$ is the infinitesimal volume element. In Cartesian coordinates $d\tau = dx\,dy\,dz$.

1.1.11 Fundamental theorem for gradients

$$\int_{\vec{a}\,\mathcal{P}}^{\vec{b}} (\nabla T) \cdot d\vec{\ell} = T(\vec{b}) - T(\vec{a})$$

1.1.12 Fundamental theorem for divergences (Gauss's theorem, Green's theorem, divergence theorem)

$$\int_{\mathcal{V}} (\nabla \cdot \vec{v})d\tau = \oint_{\mathcal{S}} \vec{v} \cdot d\vec{a}$$

1.1.13 Fundamental theorem for curls (Stoke's theorem, curl theorem)

$$\int_{\mathcal{S}} (\nabla \times \vec{v}) \cdot d\vec{a} = \oint_{\mathcal{P}} \vec{v} \cdot d\vec{\ell}$$

1.1.14 Cylindrical polar coordinates

Here our infinitesimal quantities are

$$d\vec{\ell} = ds\,\hat{s} + s\,d\phi\,\hat{\phi} + dz\,\hat{z}$$

and

$$d\tau = s\,ds\,d\phi\,dz.$$

1.1.15 Spherical polar coordinates

Here our infinitesimal quantities are

$$\mathrm{d}\vec{\ell} = \mathrm{d}r\,\hat{r} + r\,\mathrm{d}\theta\,\hat{\theta} + r\sin\theta\,\mathrm{d}\phi\,\hat{\phi}$$

and

$$\mathrm{d}\tau = r^2\sin\theta\,\mathrm{d}r\,\mathrm{d}\theta\,\mathrm{d}\phi.$$

1.1.16 One-dimensional Dirac delta function

The one-dimensional Dirac delta function is given by

$$\delta(x-a) = \begin{cases} 0 & x \neq a \\ \infty & x = a \end{cases}$$

and has the following properties

$$\int_{-\infty}^{\infty} \delta(x-a)\mathrm{d}x = 1$$

$$\int_{-\infty}^{\infty} f(x)\delta(x-a)\mathrm{d}x = f(a)$$

$$\delta(kx) = \frac{1}{|k|}\delta(x).$$

1.1.17 Theory of vector fields

If the curl of a vector field $\vec{F}$ vanishes everywhere, then $\vec{F}$ can be written as the gradient of a scalar potential V:

$$\nabla \times \vec{F} \leftrightarrow \vec{F} = -\nabla V.$$

If the divergence of a vector vanishes everywhere, then $\vec{F}$ can be expressed as the curl of a vector potential $\vec{A}$:

$$\nabla \cdot \vec{F} = 0 \leftrightarrow \vec{F} = \nabla \times \vec{A}.$$

1.2 Problems and solutions

Problem 1.1. Given vectors $\vec{A} = 3\hat{x} + 9\hat{y} + 5\hat{z}$ and $\vec{B} = \hat{x} - 7\hat{y} + 4\hat{z}$, calculate $\vec{A} \cdot \vec{B}$ and $\vec{A} \times \vec{B}$ using vector components and find the angle between $\vec{A}$ and $\vec{B}$ using both products.

Solution

$$\vec{A} \cdot \vec{B} = (3\hat{x} + 9\hat{y} + 5\hat{z}) \cdot (\hat{x} - 7\hat{y} + 4\hat{z})$$
$$= (3)(1) + (9)(-7) + (5)(4) = 3 - 63 + 20$$

$$\vec{A} \cdot \vec{B} = -40$$

$$\vec{A} \times \vec{B} = \begin{vmatrix} \hat{x} & \hat{y} & \hat{z} \\ 3 & 9 & 5 \\ 1 & -7 & 4 \end{vmatrix}$$
$$= [(9)(4) - (-7)(5)]\hat{x} + [(1)(5) - (3)(4)]\hat{y} + [(3)(-7) - (1)(9)]\hat{z}$$

$$\hat{A} \times \hat{B} = 71\hat{x} - 7\hat{y} - 30\hat{z}$$

To find the angle θ between $\vec{A}$ and $\hat{B}$ we must first calculate A and B:

$$A = \sqrt{3^2 + 9^2 + 5^2} = \sqrt{115}$$
$$B = \sqrt{1^2 + (-7)^2 + 4^2} = \sqrt{66}.$$

Using the dot product, we have

$$\vec{A} \cdot \vec{B} = AB \cos \theta \rightarrow \theta = \cos^{-1}\left(\frac{-40}{\sqrt{115}\sqrt{66}}\right)$$
$$\theta = 117.3°.$$

Using the cross product, we have

$$\left|\vec{A} \times \vec{B}\right| = AB \sin \theta \rightarrow \sqrt{71^2 + (-7)^2 + (-30)^2} = \sqrt{115}\sqrt{66} \sin \theta$$
$$\theta = 62.7°.$$

Note, however, that we can see that the angle between $\vec{A}$ and $\vec{B}$ is greater than 90°. For any argument γ, $-90° \leqslant \sin^{-1}(\gamma) \leqslant 90°$. Since the angle between $\vec{A}$ and $\vec{B}$ is greater than 90°, we must adjust for this by subtracting our angle from 180°. Therefore, $\theta = 180° - 62.7° = 117.3°$ as expected.

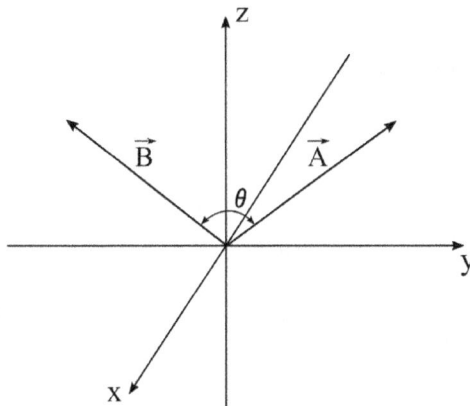

Problem 1.2. The scalar triple product states $\vec{A} \cdot (\vec{B} \times \vec{C}) = \vec{B} \cdot (\vec{C} \times \vec{A})$. Prove this by expressing each side in terms of its components.

Solution Starting with the left-hand side, the cross product is

$$
\vec{B} \times \vec{C} = \begin{vmatrix} \hat{x} & \hat{y} & \hat{z} \\ B_x & B_y & B_z \\ C_x & C_y & C_z \end{vmatrix}
$$

$$
= \left(B_y C_z - B_z C_y \right)\hat{x} + \left(B_z C_x - B_x C_z \right)\hat{y} + \left(B_x C_y - B_y C_x \right)\hat{z}.
$$

Now, dotting $\vec{A}$ with $(\vec{B} \times \vec{C})$

$$
\vec{A} \cdot \left(\vec{B} \times \vec{C} \right) = A_x\left(B_y C_z - B_z C_y \right) + A_y\left(B_z C_x - B_x C_z \right) + A_z\left(B_x C_y - B_y C_x \right)
$$

$$
= A_x B_y C_z - A_x B_z C_y + A_y B_z C_x - A_y B_x C_z + A_z B_x C_y - A_z B_y C_x
$$

$$
= B_x\left(C_y A_z - C_z A_y \right) + B_y\left(C_z A_x - C_x A_z \right) + B_z\left(C_x A_y - C_y A_x \right)
$$

$$
\vec{A} \cdot \left(\vec{B} \times \vec{C} \right) = \vec{B} \cdot \left[\left(C_y A_z - C_z A_y \right)\hat{x} + \left(C_z A_x - C_x A_z \right)\hat{y} + \left(C_x A_y - C_y A_x \right)\hat{z} \right].
$$

Note the term in brackets is precisely $\vec{C} \times \vec{A}$, therefore

$$
\vec{A} \cdot \left(\vec{B} \times \vec{C} \right) = \vec{B} \cdot \left(\vec{C} \times \vec{A} \right)
$$

as desired. This procedure can easily be applied again to prove the final part of the triple product,

$$
\vec{A} \cdot \left(\vec{B} \times \vec{C} \right) = \vec{B} \cdot \left(\vec{C} \times \vec{A} \right) = \vec{C} \cdot \left(\vec{A} \times \vec{B} \right).
$$

Problem 1.3. Given source vector $\vec{r}' = r \cos \theta\, \hat{x} + r \sin \theta\, \hat{y}$ and field vector $\vec{r} = z\hat{z}$, find the separation vector $\vec{\imath}$ and the unit vector $\hat{\imath}$.

Solution We have

$$
\vec{\imath} = \vec{r} - \vec{r}' = z\hat{z} - \left(r \cos \theta\, \hat{x} + r \sin \theta\, \hat{y} \right)
$$

$$
\vec{\imath} = -r \cos \theta\, \hat{x} - r \sin \theta\, \hat{y} + z\hat{z}.
$$

To determine the unit vector $\hat{\imath}$, we must first find the magnitude of $\vec{\imath}$,

$$
\imath = \sqrt{(-r \cos \theta)^2 + (-r \sin \theta)^2 + z^2} = \sqrt{r^2\left(\cos^2 \theta + \sin^2 \theta \right) + z^2} = \sqrt{r^2 + z^2}.
$$

So

$$\hat{r} = \frac{\vec{r}}{r} = \frac{-r\cos\theta\,\hat{x} - r\sin\theta\,\hat{y} + z\hat{z}}{\sqrt{r^2 + z^2}}.$$

Problem 1.4. Given $\vec{A}$ in coordinate system K, find the rotational matrix to give the components in system K'.

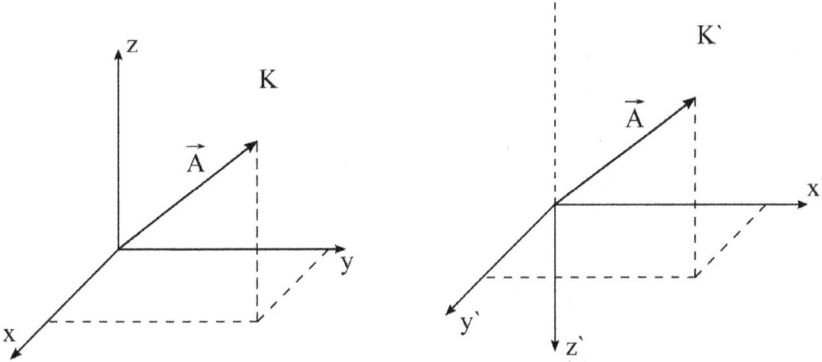

Solution From the figures, we have

$$A'_x = A_y, \qquad A'_y = A_x, \qquad A'_z = -A_z.$$

We want to find the rotational matrix R that satisfies

$$\begin{pmatrix} A'_x \\ A'_y \\ A'_z \end{pmatrix} = R \begin{pmatrix} A_x \\ A_y \\ A_z \end{pmatrix}.$$

From our equations above

$$\begin{pmatrix} A'_x \\ A'_y \\ A'_z \end{pmatrix} = \begin{pmatrix} 0 & 1 & 0 \\ 1 & 0 & 0 \\ 0 & 0 & -1 \end{pmatrix} \begin{pmatrix} A_x \\ A_y \\ A_z \end{pmatrix}.$$

Therefore,

$$R = \begin{pmatrix} 0 & 1 & 0 \\ 1 & 0 & 0 \\ 0 & 0 & -1 \end{pmatrix}.$$

Problem 1.5. Find the gradient of the following functions:
a) $T = x^4 + y^2 + z^3$
b) $T = x^2 \ln y\, z^3$
c) $T = x^2 y + z^3$

Solutions
a) $T = x^4 + y^2 + z^3$

$$\nabla T = \frac{\partial T}{\partial x}\hat{x} + \frac{\partial T}{\partial y}\hat{y} + \frac{\partial T}{\partial z}\hat{z} = 4x^3\hat{x} + 2y\hat{y} + 3z^2\hat{z}$$

b) $T = x^2 \ln y\, z^3$

$$\nabla T = \frac{\partial T}{\partial x}\hat{x} + \frac{\partial T}{\partial y}\hat{y} + \frac{\partial T}{\partial z}\hat{z} = 2xz^3 \ln y\, \hat{x} + \frac{x^2 z^3}{y}\hat{y} + 3x^2 z^2 \ln y\, \hat{z}$$

c) $T = x^2 y + z^3$

$$\nabla T = \frac{\partial T}{\partial x}\hat{x} + \frac{\partial T}{\partial y}\hat{y} + \frac{\partial T}{\partial z}\hat{z} = 2xy\hat{x} + x^2\hat{y} + 3z^2\hat{z}$$

Problem 1.6. Find the divergence of the following functions:
a) $\vec{v} = xy\hat{x} - 2y^2 z\hat{y} + z^3\hat{z}$
b) $\vec{v} = (x + y)\hat{x} + (y + z)\hat{y} + (z + x)\hat{z}$

Solutions
a) $\vec{v} = xy\hat{x} - 2y^2 z\hat{y} + z^3\hat{z}$

$$\nabla \cdot \vec{v} = \frac{\partial v_x}{\partial x} + \frac{\partial v_y}{\partial y} + \frac{\partial v_z}{\partial z} = y - 4yz + 3z^2$$

b) $\vec{v} = (x + y)\hat{x} + (y + z)\hat{y} + (z + x)\hat{z}$

$$\nabla \cdot \vec{v} = \frac{\partial v_x}{\partial x} + \frac{\partial v_y}{\partial y} + \frac{\partial v_z}{\partial z} = 1 + 1 + 1 = 3$$

Problem 1.7. Find the curl of the following functions:
a) $\vec{v} = xy\hat{x} - 2y^2 z\hat{y} + z^3\hat{z}$
b) $\vec{v} = (x + y)\hat{x} + (y + z)\hat{y} + (z + x)\hat{z}$
c) $\vec{v} = \sin x\, \hat{x} + \cos y\, \hat{y}$

Solutions

a) $\vec{v} = xy\hat{x} - 2y^2z\hat{y} + z^3\hat{z}$

$$\nabla \times \vec{v} = \left(\frac{\partial v_z}{\partial y} - \frac{\partial v_y}{\partial z} \right)\hat{x} + \left(\frac{\partial v_x}{\partial z} - \frac{\partial v_z}{\partial x} \right)\hat{y} + \left(\frac{\partial v_y}{\partial x} - \frac{\partial v_x}{\partial y} \right)\hat{z}$$

$$= \left[\frac{\partial(z^3)}{\partial y} - \frac{\partial(-2y^2z)}{\partial z} \right]\hat{x} + \left[\frac{\partial(xy)}{\partial z} - \frac{\partial(z^3)}{\partial x} \right]\hat{y}$$

$$+ \left[\frac{\partial(-2y^2z)}{\partial x} - \frac{\partial(xy)}{\partial y} \right]\hat{z}$$

$$= \left(0 + 2y^2 \right)\hat{x} + (0 - 0)\hat{y} + (0 - x)\hat{z}$$

$$\nabla \times \vec{v} = 2y^2\hat{x} - x\hat{z}$$

b) $\vec{v} = (x + y)\hat{x} + (y + z)\hat{y} + (z + x)\hat{z}$

$$\nabla \times \vec{v} = \left(\frac{\partial v_z}{\partial y} - \frac{\partial v_y}{\partial z} \right)\hat{x} + \left(\frac{\partial v_x}{\partial z} - \frac{\partial v_z}{\partial x} \right)\hat{y} + \left(\frac{\partial v_y}{\partial x} - \frac{\partial v_x}{\partial y} \right)\hat{z}$$

$$= \left[\frac{\partial(z + x)}{\partial y} - \frac{\partial(y + z)}{\partial z} \right]\hat{x} + \left[\frac{\partial(x + y)}{\partial z} - \frac{\partial(z + x)}{\partial x} \right]\hat{y}$$

$$+ \left[\frac{\partial(y + z)}{\partial x} - \frac{\partial(x + y)}{\partial y} \right]\hat{z}$$

$$\nabla \times \vec{v} = -\hat{x} - \hat{y} - \hat{z}$$

c) $\vec{v} = \sin x\ \hat{x} + \cos y\ \hat{y}$

$$\nabla \times \vec{v} = \left(\frac{\partial v_z}{\partial y} - \frac{\partial v_y}{\partial z} \right)\hat{x} + \left(\frac{\partial v_x}{\partial z} - \frac{\partial v_z}{\partial x} \right)\hat{y} + \left(\frac{\partial v_y}{\partial x} - \frac{\partial v_x}{\partial y} \right)\hat{z}$$

$$= \left[\frac{\partial(0)}{\partial y} - \frac{\partial(\cos y)}{\partial z} \right]\hat{x} + \left[\frac{\partial(\sin x)}{\partial z} - \frac{\partial(0)}{\partial x} \right]\hat{y}$$

$$+ \left[\frac{\partial(\cos y)}{\partial x} - \frac{\partial(\sin x)}{\partial y} \right]\hat{z} = 0$$

Problem 1.8. Prove $\nabla \times (\nabla T) = 0$.

Solution

$$\nabla \times \left(\nabla T \right) = \begin{vmatrix} \hat{x} & \hat{y} & \hat{z} \\ \dfrac{\partial}{\partial x} & \dfrac{\partial}{\partial y} & \dfrac{\partial}{\partial z} \\ \dfrac{\partial T}{\partial x} & \dfrac{\partial T}{\partial y} & \dfrac{\partial T}{\partial z} \end{vmatrix}$$

$$= \left[\frac{\partial}{\partial y}\left(\frac{\partial T}{\partial z}\right) - \frac{\partial}{\partial z}\left(\frac{\partial T}{\partial y}\right) \right]\hat{x} + \left[\frac{\partial}{\partial z}\left(\frac{\partial T}{\partial x}\right) - \frac{\partial}{\partial x}\left(\frac{\partial T}{\partial z}\right) \right]\hat{y}$$

$$+ \left[\frac{\partial}{\partial x}\left(\frac{\partial T}{\partial y}\right) - \frac{\partial}{\partial y}\left(\frac{\partial T}{\partial x}\right) \right]\hat{z}$$

$$\nabla \times (\nabla T) = 0.$$

Problem 1.9. Find the Laplacian of the following functions:
a) $T = x + y^2 + xz + 3$
b) $T = e^x + \sin y \cos(2z)$
c) $T = \sin x \cos y$
d) $\vec{v} = xy\hat{x} + z^2\hat{y} - 2\hat{z}$

Solutions
a) $T = x + y^2 + xz + 3$

$$\nabla^2 T = \frac{\partial^2 T}{\partial x^2} + \frac{\partial^2 T}{\partial y^2} + \frac{\partial^2 T}{\partial z^2} = 0 + 2 + 0 = 2$$

b) $T = e^x + \sin y \cos(2z)$

$$\nabla^2 T = \frac{\partial^2 T}{\partial x^2} + \frac{\partial^2 T}{\partial y^2} + \frac{\partial^2 T}{\partial z^2} = e^x - \sin y \cos(2z) - 4 \sin y \cos(2z)$$

$$= e^x - 5 \sin y \cos(2z)$$

c) $T = \sin x \cos y$

$$\nabla^2 T = \frac{\partial^2 T}{\partial x^2} + \frac{\partial^2 T}{\partial y^2} + \frac{\partial^2 T}{\partial z^2} = -\sin x \cos y - \sin x \cos y = -2 \sin x \cos y$$

d) $\vec{v} = xy\hat{x} + z^2\hat{y} - 2\hat{z}$

$$\nabla^2\vec{v} = \left(\frac{\partial^2 v_x}{\partial x^2} + \frac{\partial^2 v_x}{\partial y^2} + \frac{\partial^2 v_x}{\partial z^2}\right)\hat{x} + \left(\frac{\partial^2 v_y}{\partial x^2} + \frac{\partial^2 v_y}{\partial y^2} + \frac{\partial^2 v_y}{\partial z^2}\right)\hat{y}$$

$$+ \left(\frac{\partial^2 v_z}{\partial x^2} + \frac{\partial^2 v_z}{\partial y^2} + \frac{\partial^2 v_z}{\partial z^2}\right)\hat{z}$$

$$\nabla^2\vec{v} = (0 + 0 + 0)\hat{x} + (0 + 0 + 2)\hat{y} + (0 + 0 + 0)\hat{z} = 2\hat{y}$$

Problem 1.10. Test the divergence theorem with $\vec{v} = 2xy\hat{x} + y^2z^3\hat{y} + (x^2z - 2y)\hat{z}$ and the volume below.

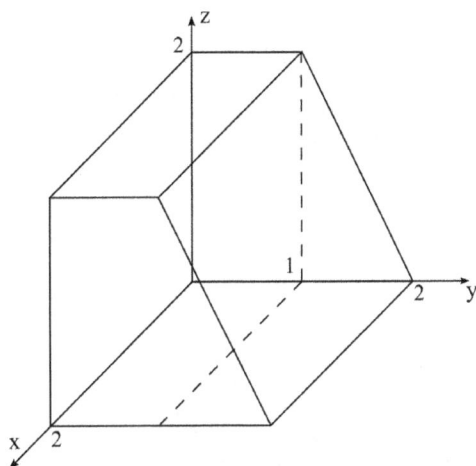

Solution The divergence theorem states

$$\int_V \nabla \cdot \vec{v}\,d\tau = \oint_S \vec{v} \cdot d\vec{a}.$$

Starting with the left-hand side, we have the divergence

$$\nabla \cdot \vec{v} = 2y + 2yz^3 + x^2 = 2y\left(z^3 + 1\right) + x^2.$$

We must split the volume into two pieces, (a) $0 \leqslant y \leqslant 1$ and (b) $1 \leqslant y \leqslant 2$.
(a)

$$\int_V \nabla \cdot \vec{v}\,d\tau = \int_0^2 \int_0^2 \int_0^1 \left[2y\left(z^3 + 1\right) + x^2\right]dy\,dx\,dz = \frac{52}{3}$$

(b)

$$\int \nabla \cdot \vec{v} d\tau = \int\limits_0^2 \int\limits_1^2 \int\limits_0^{4-2y} \left[2y\left(z^3 + 1\right) + x^2 \right] dy\ dx\ dz = \frac{176}{15}$$

So,

$$\int_v \nabla \cdot \vec{v} d\tau = \frac{52}{3} + \frac{176}{15} = \frac{436}{15}.$$

Now we solve $\oint_S \vec{v} \cdot d\vec{a}$, which must be evaluated over the six sides.

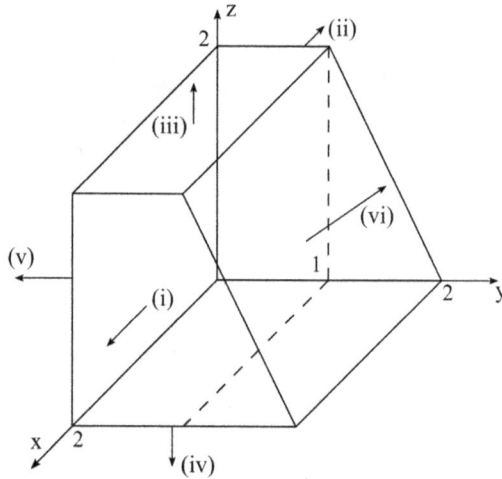

(i) We must split this region into two sections (a) and (b), and $d\vec{a} = dy\ dz\ \hat{z}$ with $x = 2$.

In (a), $0 \leqslant y \leqslant 1$,

$$\int \vec{v} \cdot d\vec{a} = \int\limits_0^2 \int\limits_0^1 2(2)y\ dy\ dz = 4.$$

In (b), $1 \leqslant y \leqslant 2$ and $0 \leqslant z \leqslant 4 - 2y$

$$\int \vec{v} \cdot d\vec{a} = \int\limits_1^2 \int\limits_0^{4-2y} 2(2)y\ dz\ dy = \frac{16}{3}.$$

(ii) Here, $d\vec{a} = -dy\, dz\, \hat{x}$ and $x = 0$, so $\vec{v} \cdot d\vec{a} = 2(0)y(-dy\, dx)\hat{x} = 0$.

(iii) Here, $d\vec{a} = dx\, dy\, \hat{z}$ and $z = 2$

$$\int \vec{v} \cdot d\vec{a} = \int_0^2 \int_0^1 \left[x^2(2) - 2y \right] dy\, dx = \frac{10}{3}.$$

(iv) Here, $d\vec{a} = -dx\, dy\, \hat{z}$ and $z = 0$

$$\int \vec{v} \cdot d\vec{a} = \int_0^2 \int_0^2 \left[x^2(0) - 2y \right](-dx\, dy) = 8.$$

(v) Here $d\vec{a} = dx\, dz\, \hat{y}$ and $y = 0$, so $\vec{v} \cdot d\vec{a} = 0^2 z^3(-dx\, dz) = 0$.

(vi) Here, we have $d\vec{a} = dx\, dz'\, \hat{n}$ where $\hat{n} = \frac{\vec{n}}{n}$. We can find $\vec{n}$ by crossing vectors
$\vec{A} = -\hat{y} + 2\hat{z}$ and $\vec{B} = 2\hat{x}$ (the edges of the volume):

$$\vec{n} = \vec{A} \times \vec{B} = \begin{vmatrix} \hat{x} & \hat{y} & \hat{z} \\ 0 & -1 & 2 \\ 2 & 0 & 0 \end{vmatrix} = 4\hat{y} + 2\hat{z}.$$

So

$$n = \sqrt{4^2 + 2^2} = 2\sqrt{5}$$

and

$$\hat{n} = \frac{2\sqrt{5}}{5}\hat{y} + \frac{\sqrt{5}}{5}\hat{z}.$$

We can also obtain dz' by considering

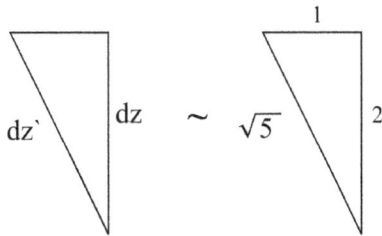

so $dz' = \frac{\sqrt{5}}{2}dz$. Now

$$d\vec{a} = \frac{\sqrt{5}}{2}dx\,dz\left(\frac{2\sqrt{5}}{5}\hat{y} + \frac{\sqrt{5}}{5}\hat{z}\right) = \left(\hat{y} + \frac{1}{2}\hat{z}\right)dx\,dz$$

and

$$z = 4 - 2y \rightarrow y = 2 - \frac{z}{2}.$$

So

$$\int \vec{v}\cdot d\vec{a} = \int_0^2\int_0^2\left[y^2z^3 + \frac{1}{2}\left(x^2z - 2y\right)\right]dx\,dz$$

$$= \int_0^2\int_0^2\left\{\left(2 - \frac{z}{2}\right)^2 z^3 + \frac{1}{2}\left[x^2z - 2\left(2 - \frac{z}{2}\right)\right]\right\}dx\,dz = \frac{42}{5}.$$

Therefore

$$\oint_S \vec{v}\cdot d\vec{a} = 4 + \frac{16}{3} + \frac{10}{3} + 8 + \frac{42}{5} = \frac{436}{15}$$

as expected.

Problem 1.11. Test the curl theorem with $\vec{v} = 5xy^2\hat{x} + yz^2\hat{y} + 4x^2z\hat{z}$ and the surface below.

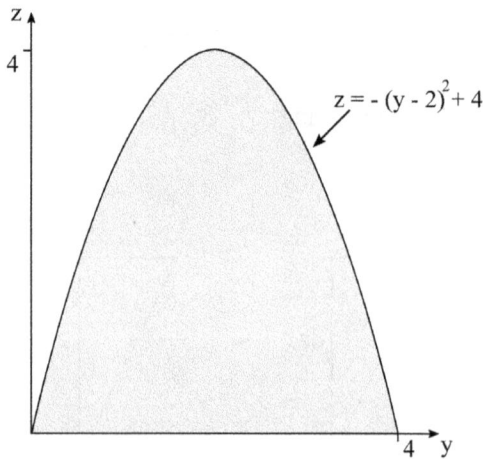

$$z = -(y-2)^2 + 4$$

Solution The curl theorem states

$$\int_{S} (\nabla \times \vec{v}) \cdot d\vec{a} = \oint_{\mathcal{P}} \vec{v} \cdot d\vec{\ell}.$$

Starting with the left-hand side, the curl is given by

$$\nabla \times \vec{v} = \begin{vmatrix} \hat{x} & \hat{y} & \hat{z} \\ \dfrac{\partial}{\partial x} & \dfrac{\partial}{\partial y} & \dfrac{\partial}{\partial z} \\ 5xy^2 & yz^2 & 4x^2z \end{vmatrix} = -2yz\hat{x} - 8xz\hat{y} - 10xy\hat{z}$$

We also have $d\vec{a} = dy\, dz\, \hat{x}$ with $0 \leqslant z \leqslant -(y-2)^2 + 4$. So

$$\int_{S} (\nabla \times \vec{v}) \cdot d\vec{a} = \int_{0}^{4} \int_{0}^{-(y-2)^2+4} -2yz\, dz\, dy = -\frac{1024}{15}.$$

Now to solve $\oint_{\mathcal{P}} \vec{v} \cdot d\vec{\ell}$ over the two paths (i) and (ii):

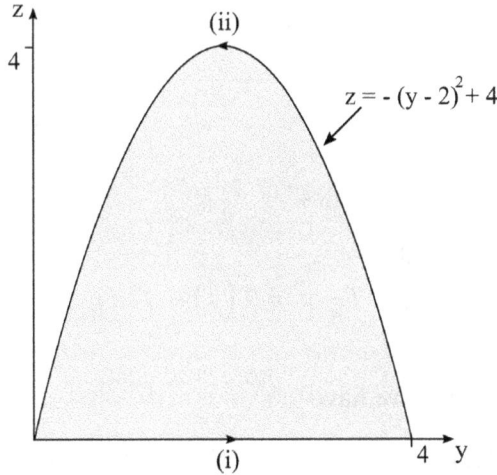

(i) Here we have $x = 0$, $z = 0$, and $d\vec{\ell} = dy\, \hat{y}$. So $\vec{v} \cdot d\vec{\ell} = y(0^2)dy = 0$.

(ii) Here we have $d\vec{\ell} = dy\, \hat{y} + dz\, \hat{z}$, $x = 0$, and $z = -(y-2)^2 + 4$

$$\int_{\mathcal{P}} \vec{v} \cdot d\vec{\ell} = \int_{\mathcal{P}} yz^2 dy + 4(0^2)z\, dz = \int_{4}^{0} y\left[-(y-2)^2 + 4\right]^2 dy = -\frac{1024}{15}.$$

So,

$$\oint_{\mathcal{P}} \vec{v} \cdot d\vec{\ell} = 0 + \frac{-1024}{15} = \frac{-1024}{15}$$

as expected.

Problem 1.12. Test the gradient theorem with $T = 3xz^2 - y^2z$ and path $z = y^2$ and $z = y^3$ from $(0, 0, 0) \rightarrow (0, 1, 1)$.

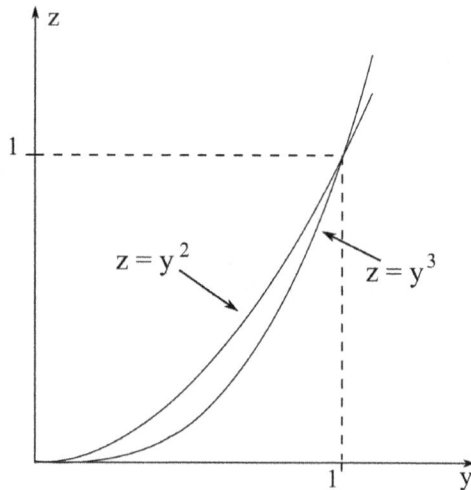

Solution The gradient theorem states

$$\int_{\vec{a}}^{\vec{b}} \nabla T \cdot d\vec{\ell} = T(\vec{b}) - T(\vec{a}).$$

Starting with the right side, we have

$$T(0,1,1) - T(0,0,0) = 3(0)(1^2) - (1^2)(1) - 0 = -1.$$

Now to solve $\int_{\vec{a}}^{\vec{b}} \nabla T \cdot d\vec{\ell}$, the gradient of T is given by

$$\nabla T = \frac{\partial}{\partial x}\left(3xz^2 - y^2z\right)\hat{x} + \frac{\partial}{\partial y}\left(3xz^2 - y^2z\right)\hat{y} + \frac{\partial}{\partial z}\left(3xz^2 - y^2z\right)\hat{z}$$

$$\nabla T = 3z^2\hat{x} - 2yz\,\hat{y} + \left(6xz - y^2\right)\hat{z}.$$

Here, $d\vec{\ell} = dy\,\hat{y} + dz\,\hat{z}$ with $x = 0$. So

$$\nabla T \cdot d\vec{\ell} = -2yz\,dy + \left[6(0)(z) - y^2\right]dz = -2yz\,dy - y^2dz.$$

For path (i), we have $z = y^2 \rightarrow dz = 2y\,dy$. So

$$\nabla T \cdot d\vec{\ell} = -2y\left(y^2\right)dy - y^2(2y\,dy) = -4y^3\,dy$$

and

$$\int_{\vec{a}}^{\vec{b}} \nabla T \cdot d\vec{\ell} = \int_0^1 -4y^3 dy = -1$$

as expected. For path (ii), we have $z = y^3 \rightarrow dz = 3y^2dy$. So

$$\nabla T \cdot d\vec{\ell} = -2y\left(y^3\right)dy - y^2\left(3y^2dy\right) = -5y^4dy$$

and

$$\int_{\vec{a}}^{\vec{b}} \nabla T \cdot d\vec{\ell} = \int_0^1 -5y^4 dy = -1$$

also as expected.

Problem 1.13. Verify the following integration by parts given $f = xy^2z$ and $\vec{A} = z^2\hat{x} + 4xy\hat{y} - x^2z\hat{z}$ and the surface below,

$$\int_S f\left(\nabla \times \vec{A}\right) \cdot d\vec{a} = \int_S \left[\vec{A} \times \left(\nabla f\right)\right] \cdot d\vec{a} + \oint_P f\vec{A} \cdot d\vec{\ell}.$$

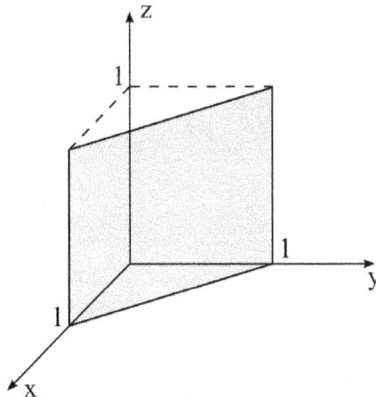

Solution Starting with the left-hand side

$$\nabla \times \vec{A} = \begin{vmatrix} \hat{x} & \hat{y} & \hat{z} \\ \dfrac{\partial}{\partial x} & \dfrac{\partial}{\partial y} & \dfrac{\partial}{\partial z} \\ z^2 & 4xy & -x^2z \end{vmatrix} = [2z - (-2xz)]\hat{y} + 4y\hat{z} = 2z(x + 1)\hat{y} + 4y\hat{z}.$$

Now

$$f\left(\nabla \times \vec{A}\right) = 2xy^2z^2(x + 1)\hat{y} + 4xy^3z\hat{z}.$$

Here we have $d\vec{a} = dx'\, dz\, \hat{n}$ where $\hat{n} = \hat{x} + \hat{y}$ and $n = \sqrt{2}$ so $\hat{n} = \frac{\sqrt{2}}{2}\hat{x} + \frac{\sqrt{2}}{2}\hat{y}$. Also from

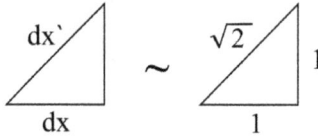

we have

$$dx' = \sqrt{2}\, dx \quad \text{with } y = 1 - x.$$

Now

$$d\vec{a} = \sqrt{2}\, dx\, dz\left(\frac{\sqrt{2}}{2}\hat{x} + \frac{\sqrt{2}}{2}\hat{y}\right) = dx\, dz\left(\hat{x} + \hat{y}\right).$$

Therefore,

$$\int_S f\left(\nabla \times \vec{A}\right) \cdot d\vec{a} = \int_0^1 \int_0^1 \left[2xy^2z^2(x + 1)\hat{y} + 4xy^3z\hat{z}\right] \cdot (\hat{x} + \hat{y})dx\, dz$$

$$= \int_0^1 \int_0^1 2x(1 - x)^2z^2(x + 1)dx\, dz = \frac{7}{90}.$$

Next, we will solve the $\oint_P f\vec{A} \cdot d\vec{\ell}$ term for the four segments.

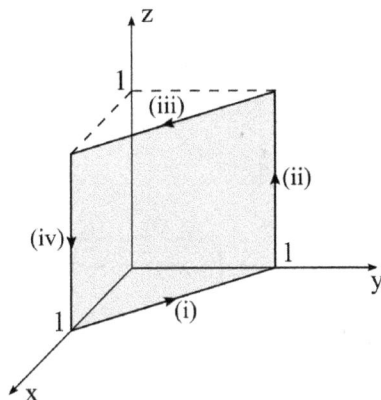

Segment (i)

$$z = 0 \rightarrow f = xy^2(0) = 0.$$

Segment (ii)

$$x = 0 \rightarrow f = (0)y^2z = 0.$$

Segment (iii)

$$d\vec{l} = dx\,\hat{x} + dy\,\hat{y}, z = 1, \qquad \text{and} \qquad y = 1 - x \rightarrow dy = -dx.$$

Segment (iv)

$$y = 0 \rightarrow f = x(0^2)z = 0.$$

So

$$f\left(\vec{A} \cdot d\vec{l}\right) = xy^2\left(z^2dx + 4xy\,dy\right) = x(1 - x)^2\left[1 - 4x(1 - x)\right]dx$$

and

$$\oint_{\mathcal{P}} f\vec{A} \cdot d\vec{l} = \int_0^1 x(1 - x)^2\left[1 - 4x(1 - x)\right]dx = \frac{1}{60}.$$

Now to solve the $\int_{S} [A \times \nabla f] \cdot d\vec{a}$ term. First, we have

$$\nabla f = y^2z\hat{x} + 2xyz\hat{y} + xy^2\hat{z}.$$

So

$$A \times (\nabla f) = \begin{vmatrix} \hat{x} & \hat{y} & \hat{z} \\ z^2 & 4xy & -x^2z \\ y^2z & 2xyz & xy^2 \end{vmatrix}$$

$$= \left(4x^2y^3 + 2x^3yz^2\right)\hat{x} + \left(-x^2y^2z^2 - xy^2z^2\right)\hat{y} + \left(2xyz^3 - 4xy^3z\right)\hat{z}.$$

As before, $d\vec{a} = dx\, dz(\hat{x} + \hat{y})$. So

$$\int_S [A \times (\nabla f)] \cdot d\vec{a} = \int_0^1 \int_0^1 \left[4x^2(1-x)^3 + 2x^3(1-x)z^2 \right.$$
$$\left. -x^2(1-x)^2z^2 - x(1-x)^2z^2\right] dx\, dz$$

$$\int_S [A \times (\nabla f)] \cdot d\vec{a} = \frac{11}{180}.$$

So

$$\int_S [A \times (\nabla f)] \cdot d\vec{a} + \oint_P f\vec{A} \cdot d\vec{\ell} = \frac{11}{180} + \frac{1}{60} = \frac{7}{90}$$

as expected.

Problem 1.14. Find the divergence and curl of the following functions:
a) $\vec{v} = r^2\hat{r} + \cos\theta \sin\phi\, \hat{\theta} + \sin\theta \cos\phi\, \hat{\phi}$
b) $\vec{v} = s\cos\phi\, \hat{s} + \cos\phi \sin\phi\, \hat{\phi} + z\sin\phi\, \hat{z}$

Solutions
a) $\vec{v} = r^2\hat{r} + \cos\theta \sin\phi\, \hat{\theta} + \sin\theta \cos\phi\, \hat{\phi}$

$$\nabla \cdot \vec{v} = \frac{1}{r^2}\frac{\partial}{\partial r}\left(r^2 v_r\right) + \frac{1}{r\sin\theta}\frac{\partial}{\partial\theta}\left(\sin\theta\, v_\theta\right) + \frac{1}{r\sin\theta}\frac{\partial v_\phi}{\partial\phi}$$

$$= \frac{1}{r^2}\frac{\partial}{\partial r}\left(r^4\right) + \frac{1}{r\sin\theta}\frac{\partial}{\partial\theta}\left(\sin\theta \cos\theta \sin\phi\right) + \frac{1}{r\sin\theta}\frac{\partial}{\partial\phi}\left(\sin\theta \cos\phi\right)$$

$$= \frac{1}{r^2}\left(4r^3\right) + \frac{\sin\phi}{r\sin\theta}\left(-\sin^2\theta + \cos^2\theta\right) + \frac{1}{r}(-\sin\phi)$$

$$= 4r + \frac{\sin\phi}{r\sin\theta}\left(1 - 2\sin^2\theta\right) - \frac{\sin\phi}{r}$$

$$\nabla \cdot \vec{v} = 4r + \frac{\sin\phi}{r}\left(\csc\theta - 2\sin\theta - 1\right)$$

$$\nabla \times \vec{v} = \frac{1}{r\sin\theta}\left[\frac{\partial}{\partial\theta}\left(\sin\theta\, v_\phi\right) - \frac{\partial v_\theta}{\partial\phi}\right]\hat{r}$$

$$+ \frac{1}{r}\left[\frac{1}{\sin\theta}\frac{\partial v_r}{\partial\phi} - \frac{\partial}{\partial r}(rv_\phi)\right]\hat{\theta} + \frac{1}{r}\left[\frac{\partial}{\partial r}(rv_\theta) - \frac{\partial v_r}{\partial\theta}\right]\hat{\phi}$$

$$= \frac{1}{r\sin\theta}\left[\frac{\partial}{\partial\theta}\left(\sin^2\theta\cos\phi\right) - \frac{\partial}{\partial\phi}(\cos\theta\sin\phi)\right]\hat{r}$$

$$+ \frac{1}{r}\left[\frac{1}{\sin\theta}\frac{\partial}{\partial\phi}\left(r^2\right) - \frac{\partial}{\partial r}(r\sin\theta\cos\phi)\right]\hat{\theta}$$

$$+ \frac{1}{r}\left[\frac{\partial}{\partial r}(r\cos\theta\sin\phi) - \frac{\partial}{\partial\phi}\left(r^2\right)\right]\hat{\phi}$$

$$= \frac{1}{r\sin\theta}\left(2\sin\theta\cos\theta\cos\phi - \cos\theta\cos\phi\right)\hat{r} - \frac{\sin\theta\cos\phi}{r}\hat{\theta}$$
$$+ \frac{\cos\theta\sin\phi}{r}\hat{\phi}$$

$$\nabla \times \vec{v} = \frac{\cos\theta\cos\phi}{r}(2 - \csc\theta)\hat{r} - \frac{\sin\theta\cos\phi}{r}\hat{\theta} + \frac{\cos\theta\sin\phi}{r}\hat{\phi}$$

b) $\vec{v} = s\cos\phi\,\hat{s} + \cos\phi\sin\phi\,\hat{\phi} + z\sin\phi\,\hat{z}$

$$\nabla \cdot \vec{v} = \frac{1}{s}\frac{\partial}{\partial s}(sv_s) + \frac{1}{s}\frac{\partial v_\phi}{\partial\phi} + \frac{\partial v_z}{\partial z}$$

$$= \frac{1}{s}\frac{\partial}{\partial s}\left(s^2\cos\phi\right) + \frac{1}{s}\frac{\partial}{\partial\phi}(\cos\phi\sin\phi) + \frac{\partial}{\partial z}(z\sin\phi)$$

$$= 2\cos\phi + \frac{1}{s}\left(-\sin^2\phi + \cos^2\phi\right) + \sin\phi$$

$$\nabla \cdot \vec{v} = 2\cos\phi + \sin\phi + \frac{\cos^2\phi - \sin^2\phi}{s}$$

$$\nabla \times \vec{v} = \left(\frac{1}{s} \frac{\partial v_z}{\partial \phi} - \frac{\partial v_\phi}{\partial z} \right) \hat{s} + \left(\frac{\partial v_s}{\partial z} - \frac{\partial v_z}{\partial s} \right) \hat{\phi} + \frac{1}{s} \left[\frac{\partial}{\partial s} (s v_\phi) - \frac{\partial v_s}{\partial \phi} \right] \hat{z}$$

$$= \left[\frac{1}{s} \frac{\partial}{\partial \phi} (z \sin \phi) - \frac{\partial}{\partial z} (\cos \phi \sin \phi) \right] \hat{s} + \left[\frac{\partial}{\partial z} (s \cos \phi) - \frac{\partial}{\partial s} (z \sin \phi) \right] \hat{\phi}$$

$$+ \frac{1}{s} \left[\frac{\partial}{\partial s} (s \cos \phi \sin \phi) - \frac{\partial}{\partial \phi} (s \cos \phi) \right] \hat{z}$$

$$= \frac{z}{s} \cos \phi \, \hat{s} + \frac{1}{s} (\cos \phi \sin \phi + s \sin \phi) \hat{z}$$

$$\nabla \times \vec{v} = \frac{z}{s} \cos \phi \, \hat{s} + \frac{\sin \phi}{s} (\cos \phi + s) \hat{z}$$

Problem 1.15. Find the gradient and Laplacian of:
a) $T = r^2 (\cos \theta \sin \phi + \sin \theta \cos \phi)$
b) $T = z^2 \sin \phi - s \cos^2 \phi$

Solutions
a) $T = r^2 (\cos \theta \sin \phi + \sin \theta \cos \phi)$

$$\nabla T = \frac{\partial T}{\partial r} \hat{r} + \frac{1}{r} \frac{\partial T}{\partial \theta} \hat{\theta} + \frac{1}{r \sin \theta} \frac{\partial T}{\partial \phi} \hat{\phi}$$

$$= 2r (\cos \theta \sin \phi + \sin \theta \cos \phi) \hat{r} + \frac{1}{r} r^2 (-\sin \theta \sin \phi + \cos \theta \cos \phi) \hat{\theta}$$

$$+ \frac{1}{r \sin \theta} r^2 (\cos \theta \cos \phi - \sin \theta \sin \phi) \hat{\phi}$$

$$= 2r (\cos \theta \sin \phi + \sin \theta \cos \phi) \hat{r} + r (\cos \theta \cos \phi - \sin \theta \sin \phi) \hat{\theta}$$

$$+ \frac{r}{\sin \theta} (\cos \theta \cos \phi - \sin \theta \sin \phi) \hat{\phi}$$

$$\nabla T = 2r \sin(\theta + \phi) \hat{r} + r \cos(\theta + \phi) \hat{\theta} + \frac{r}{\sin \theta} \cos(\theta + \phi) \hat{\phi}.$$

Note we could have written T as $T = r^2 \sin(\theta + \phi)$ and then computed the gradient.

$$\nabla^2 T = \frac{1}{r^2} \frac{\partial}{\partial r} \left(r^2 \frac{\partial T}{\partial r} \right) + \frac{1}{r^2 \sin \theta} \frac{\partial}{\partial \theta} \left(\sin \theta \frac{\partial T}{\partial \theta} \right) + \frac{1}{r^2 \sin^2 \theta} \left(\frac{\partial^2 T}{\partial \phi^2} \right)$$

$$= \frac{1}{r^2}\frac{\partial}{\partial r}\left[2r^3 \sin(\theta + \phi)\right] + \frac{1}{r^2 \sin\theta}\frac{\partial}{\partial\theta}\left[r^2 \sin\theta \cos(\theta + \phi)\right]$$

$$+ \frac{1}{r^2 \sin^2\theta}\frac{\partial}{\partial\phi}\left[r^2 \cos(\theta + \phi)\right]$$

$$= 6\sin(\theta + \phi) + \frac{1}{\sin\theta}[\cos\theta\cos(\theta + \phi) - \sin\theta\sin(\theta + \phi)]$$

$$+ \frac{1}{\sin^2\theta}\left(-\sin(\theta + \phi)\right)$$

$$\nabla^2 T = 5\sin(\theta + \phi) + \frac{\cos\theta}{\sin\theta}\cos(\theta + \phi) - \frac{\sin(\theta + \phi)}{\sin^2\theta}.$$

b) $T = z^2 \sin\phi - s\cos^2\phi$

$$\nabla T = \frac{\partial T}{\partial s}\hat{s} + \frac{1}{s}\frac{\partial T}{\partial\phi}\hat{\phi} + \frac{\partial T}{\partial z}\hat{z}$$

$$= \frac{\partial}{\partial s}\left(z^2 \sin\phi - s\cos^2\phi\right)\hat{s} + \frac{1}{s}\frac{\partial}{\partial\phi}\left(z^2 \sin\phi - s\cos^2\phi\right)\hat{\phi}$$

$$+ \frac{\partial}{\partial z}\left(z^2 \sin\phi - s\cos^2\phi\right)\hat{z}$$

$$= -\cos^2\phi\,\hat{s} + \frac{1}{s}\left[z^2 \cos\phi - 2s\cos\phi(-\sin\phi)\right]\hat{\phi} + 2z\sin\phi\,\hat{z}$$

$$\nabla T = -\cos^2\phi\,\hat{s} + \frac{\cos\phi}{s}\left(z^2 + 2s\sin\phi\right)\hat{\phi} + 2z\sin\phi\,\hat{z}$$

$$\nabla^2 T = \frac{1}{s}\frac{\partial}{\partial s}\left(s\frac{\partial T}{\partial s}\right) + \frac{1}{s^2}\frac{\partial^2 T}{\partial\phi^2} + \frac{\partial^2 T}{\partial z^2}$$

$$\frac{\partial T}{\partial s} = -\cos^2\phi \rightarrow s\frac{\partial T}{\partial s} = -s\cos^2\phi \rightarrow \frac{\partial}{\partial s}\left(s\frac{\partial T}{\partial s}\right) = -\cos^2\phi$$

$$\frac{\partial T}{\partial\phi} = z^2 \cos\phi + 2s\cos\phi\sin\phi \rightarrow \frac{\partial^2 T}{\partial\phi^2} = -z^2 \sin\phi + 2s\left(-\sin^2\phi + \cos^2\phi\right)$$

$$\frac{\partial T}{\partial z} = 2z\sin\phi \rightarrow \frac{\partial^2 T}{\partial z^2} = 2\sin\phi$$

$$\nabla^2 T = -\frac{\cos^2\phi}{s} + \frac{2}{s}\left(-\sin^2\phi + \cos^2\phi\right) - \frac{z^2}{s^2}\sin\phi + 2\sin\phi$$

$$\nabla^2 T = \frac{\cos^2 \phi}{s} - \frac{2}{s}\sin^2 \phi + \left(2 - \frac{z^2}{s^2}\right)\sin \phi$$

Problem 1.16. Test the divergence theorem with $\vec{v} = r\cos\phi\,\hat{r} + r\cos\theta\sin\theta\,\hat{\theta} + r\sin\phi\,\hat{\phi}$ and the volume below (the upper half of the sphere of radius R with a cone of radius $a = \frac{R}{\sqrt{3}}$ cut out).

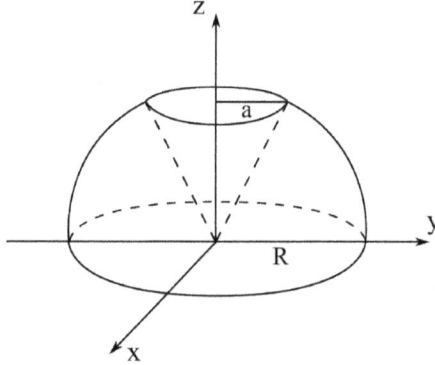

Solution The divergence theorem states

$$\int_V \nabla \cdot \vec{v}\,d\tau = \oint_S \vec{v} \cdot d\vec{a}.$$

Starting with the left-hand side, the divergence is

$$\nabla \cdot \vec{v} = \frac{1}{r^2}\frac{\partial}{\partial r}\left(r^2 v_r\right) + \frac{1}{r\sin\theta}\frac{\partial}{\partial\theta}\left(\sin\theta\,v_\theta\right) + \frac{1}{r\sin\theta}\frac{\partial v_\phi}{\partial\phi}$$

$$= \frac{1}{r^2}\frac{\partial}{\partial r}\left(r^3\cos\phi\right) + \frac{1}{r\sin\theta}\frac{\partial}{\partial\theta}\left(r\sin^2\theta\cos\theta\right) + \frac{1}{r\sin\theta}\frac{\partial}{\partial\phi}(r\sin\phi)$$

$$= 3\cos\phi + \frac{1}{\sin\theta}\left(2\sin\theta\cos^2\theta - \sin^3\theta\right) + \frac{\cos\phi}{\sin\theta}$$

$$\nabla \cdot \vec{v} = 3\cos\phi + 2\cos^2\theta - \sin^2\theta + \frac{\cos\phi}{\sin\theta}.$$

For the volume,

$$0 \leqslant r \leqslant R, \quad \tan^{-1}\left(\frac{a}{R}\right) = \tan^{-1}\left(\frac{1}{\sqrt{3}}\right) = \frac{\pi}{6} \rightarrow \frac{\pi}{6} \leqslant \theta \leqslant \frac{\pi}{2}, \quad 0 \leqslant \phi \leqslant 2\pi.$$

So

$$\int_{\mathcal{V}} \nabla \cdot \vec{v} \, d\tau = \int_0^R \int_{\frac{\pi}{6}}^{\frac{\pi}{2}} \int_0^{2\pi} \left(3\cos\phi + 2\cos^2\theta - \sin^2\theta + \frac{\cos\phi}{\sin\theta} \right) (r^2\sin\theta) d\phi \, d\theta \, dr$$

$$\int_{\mathcal{V}} \nabla \cdot \vec{v} d\tau = -\frac{\sqrt{3}}{12}\pi R^3.$$

Now for the right-hand side, we have three surfaces: the bottom (i), the outer shell (ii), and the inner part where the cone is cut out (iii). We have

$$\vec{v} = r\cos\phi \, \hat{r} + r\sin\phi \, \hat{\phi} + r\cos\theta\sin\theta \, \hat{\theta}.$$

For (i), we have $d\vec{a} = r \, dr \, d\phi \, \hat{\theta}$ and $\theta = \frac{\pi}{2}$. So

$$\vec{v} \cdot d\vec{a} = r^2 \cos\frac{\pi}{2} \sin\frac{\pi}{2} dr \, d\phi = 0$$

and

$$\int_{(i)} \vec{v} \cdot d\vec{a} = 0.$$

For (ii), we have $r = R$ and $d\vec{a} = r^2 \sin\theta \, d\theta \, d\phi \, \hat{r} = R^2 \sin\theta \, d\theta \, d\phi \, \hat{r}$. So

$$\vec{v} \cdot d\vec{a} = R^3 \cos\phi \sin\theta \, d\theta \, d\phi$$

and

$$\int_{(ii)} \vec{v} \cdot d\vec{a} = R^3 \int_0^{2\pi} \int_{\frac{\pi}{6}}^{\frac{\pi}{2}} \cos\phi \sin\theta \, d\theta \, d\phi = 0.$$

For (iii), we have $\theta = \frac{\pi}{6}$ and $d\vec{a} = -r\sin\theta \, dr \, d\phi \, \hat{\theta} = -\frac{1}{2}r \, dr \, d\phi \, \hat{\theta}$. So

$$\vec{v} \cdot d\vec{a} = -\frac{1}{2}r^2 \cos\frac{\pi}{6} \sin\frac{\pi}{6} = -\frac{\sqrt{3}}{8}r^2$$

and

$$\int_{(iii)} \vec{v} \cdot d\vec{a} = -\frac{\sqrt{3}}{8} \int_0^R \int_0^{2\pi} r^2 d\phi \, dr = -\frac{\sqrt{3}}{12}\pi R^3.$$

Therefore,

$$\oint_S \vec{v} \cdot d\vec{a} = -\frac{\sqrt{3}}{12}\pi R^3$$

as expected.

Problem 1.17. Test the curl theorem with $\vec{v} = s^2 z\, \hat{s} + \sin\phi \cos\phi\, \hat{\phi} + zs \cos\phi\, \hat{z}$ and half of a cylindrical shell with radius R and height h.

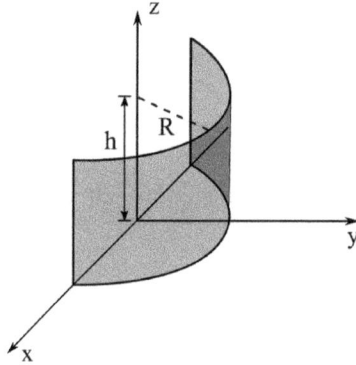

Solution The curl theorem states

$$\int_S \left(\nabla \times \vec{v}\right) \cdot d\vec{a} = \oint_p \vec{v} \cdot d\vec{\ell}.$$

Starting with the left-handed side, we have

$$d\vec{a} = s\, d\phi\, dz\, \hat{s} = R\, d\phi\, dz\, \hat{s}.$$

Since we are dotting $d\vec{a}$ with $\nabla \times \vec{v}$, we only need the $\hat{s}$ component of the curl:

$$[\nabla \times \vec{v}]_s = \left(\frac{1}{s}\frac{\partial v_z}{\partial \phi} - \frac{\partial v_\phi}{\partial z}\right)\hat{s} = \left[\frac{1}{s}\frac{\partial}{\partial \phi}(zs\cos\phi) - \frac{\partial}{\partial z}(\sin\phi\cos\phi)\right]\hat{s}$$

$$= -z\sin\phi\, \hat{s}.$$

So

$$\left(\nabla \times \vec{v}\right) \cdot d\vec{a} = -Rz\sin\phi\, d\phi\, dz.$$

We have

$$0 \leqslant \phi \leqslant \pi \qquad \text{and} \qquad 0 \leqslant z \leqslant h$$

so

$$\int_S \left(\nabla \times \vec{v}\right) \cdot d\vec{a} = \int_0^h \int_0^\pi -Rz \sin\phi \, d\phi \, dz = -h^2 R.$$

For the left-hand side, we have four curves

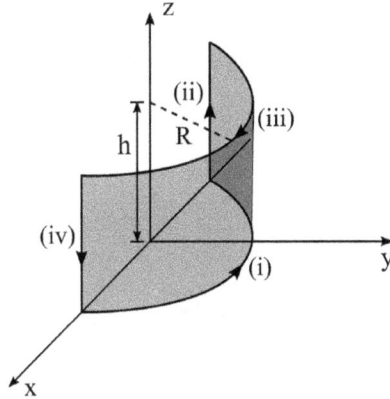

with

$$\vec{v} = s^2 z \, \hat{s} + \sin\phi \cos\phi \, \hat{\phi} + z \cos\phi \, \hat{z}.$$

For curve (i), $d\vec{\ell} = d\phi \, \hat{\phi}$, $z = 0$, and $s = R$. So

$$\vec{v} \cdot d\vec{\ell} = \sin\phi \cos\phi \, d\phi$$

and

$$\int_0^\pi \sin\phi \cos\phi \, d\phi = 0.$$

For curve (ii), $d\vec{\ell} = dz \, \hat{z}$, $\phi = \pi$, and $s = R$. So

$$\vec{v} \cdot d\vec{\ell} = zs \cos\phi \, dz = zR \cos\pi \, dz = -zR \, dz$$

and

$$\int_0^h -zR \, dz = -\frac{1}{2}h^2 R.$$

For curve (iii), $d\vec{\ell} = d\phi\,\hat{\phi}$, $z = h$, and $s = R$. So

$$\vec{v} \cdot d\vec{\ell} = \sin\phi\cos\phi\,d\phi$$

and

$$\int_{\pi}^{0} \sin\phi\cos\phi\,d\phi = 0.$$

For curve (iv), $d\vec{\ell} = dz\,\hat{z}$, $\phi = 0$, and $s = R$. So

$$\vec{v} \cdot d\vec{\ell} = zs\cos\phi\,dz = zR\cos(0)dz = zR\,dz$$

and

$$\int_{h}^{0} zR\,dz = -\frac{1}{2}h^2R^2.$$

So,

$$\oint_{\mathcal{P}} \vec{v} \cdot d\vec{\ell} = -\frac{1}{2}h^2R - \frac{1}{2}h^2R = -h^2R$$

as expected.

Problem 1.18. Test the gradient theorem using $T = sz^2 \sin\phi$ and the half helix path (radius R, height h).

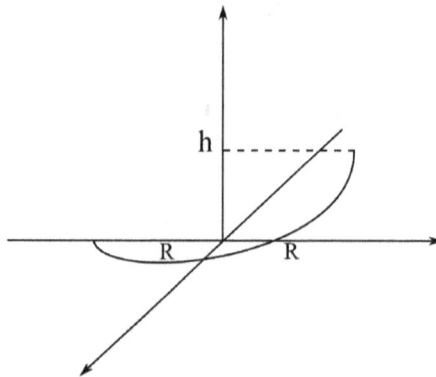

Solution The gradient theorem states

$$\int_{\mathcal{P}} \nabla T \cdot d\vec{\ell} = T(\vec{b}) - T(\vec{a}).$$

Starting with the right-hand side

$$T(\vec{b}) - T(\vec{a}) = T\left(R, \frac{\pi}{2}, h\right) - T\left(R, -\frac{\pi}{2}, 0\right) = Rh^2 \sin\frac{\pi}{2} - R(0)^2 \sin\left(-\frac{\pi}{2}\right) = h^2 R.$$

Now, the gradient is

$$\nabla T = \frac{\partial T}{\partial s}\hat{s} + \frac{1}{s}\frac{\partial T}{\partial \phi}\hat{\phi} + \frac{\partial T}{\partial z}\hat{z} = z^2 \sin\phi\,\hat{s} + z^2 \cos\phi\,\hat{\phi} + 2sz \sin\phi\,\hat{z}.$$

We also have $s = R$ and $\vec{\ell} = s\,d\phi\,\hat{\phi} + dz\,\hat{z} = R\,d\phi\,\hat{\phi} + dz\,\hat{z}$. So

$$\nabla T \cdot d\vec{\ell} = Rz^2 \cos\phi\,d\phi + 2Rz \sin\phi\,dz.$$

We need a way to relate z and ϕ. Note that as ϕ increases, z increases linearly. So, using the equation of line

$$z - z_0 = \gamma(\phi - \phi_0),$$

when $z = 0$ and $\phi = -\frac{\pi}{2}$,

$$z = \gamma\left(\phi + \frac{\pi}{2}\right),$$

when $z = h$ and $\phi = \frac{\pi}{2}$,

$$h = \gamma\left(\frac{\pi}{2} + \frac{\pi}{2}\right) \rightarrow \gamma = \frac{h}{\pi},$$

so

$$z = \frac{h}{\pi}\phi - \frac{h}{2}$$

and

$$dz = \frac{h}{\pi}d\phi.$$

Using our expressions for z and dz, we have

$$\nabla T \cdot d\vec{\ell} = \left[R\left(\frac{h}{\pi}\phi + \frac{h}{2}\right)^2 \cos\phi + 2R\left(\frac{h}{\pi}\phi + \frac{h}{2}\right)\sin\phi\left(\frac{h}{\pi}\right)\right]d\phi.$$

So

$$\int_{\vec{a}}^{\vec{b}} \nabla T \cdot d\vec{\ell} = \int_{-\frac{\pi}{2}}^{\frac{\pi}{2}} \left[R\left(\frac{h}{\pi}\phi + \frac{h}{2}\right)^2 \cos\phi + 2R\left(\frac{h}{\pi}\phi + \frac{h}{2}\right)\sin\phi\left(\frac{h}{\pi}\right)\right]d\phi = h^2 R$$

as expected.

Problem 1.19. Evaluate the following integrals:

a) $\displaystyle\int_1^3 (2x^2 - x + 4)\delta(x - 2)dx$

b) $\displaystyle\int_{-1}^1 (x^2 + 4)\delta(x - 2)dx$

c) $\displaystyle\int_2^6 \sin(\tfrac{3x}{2})\delta(x - \pi)dx$

d) $\displaystyle\int_{-2}^2 (2x^3 + 1)\delta(4x)dx$

e) $\displaystyle\int_{-\infty}^\infty x^2\delta(2x + 1)dx$

f) $\displaystyle\int_0^a \delta(x - b)dx$

Solutions

a)

$$\int_1^3 (2x^2 - x + 4)\delta(x - 2)dx.$$

Since $2 \in (1, 3)$ and $f(x) = 2x^2 - x + 4$, we have

$$\int_1^3 (2x^2 - x + 4)\delta(x - 2)dx = f(2) = 2(2)^2 - 2 + 4 = 10.$$

b)

$$\int_{-1}^1 (x^2 + 4)\delta(x - 2)dx.$$

Since $2 \notin (-1, 1)$, we have

$$\int_{-1}^1 (x^2 + 4)\delta(x - 2)dx = 0.$$

c)

$$\int_2^6 \sin\left(\frac{3x}{2}\right)\delta(x - \pi)dx.$$

Since $\pi \in (2, 6)$ and $f(x) = \sin(\frac{3x}{2})$, we have

$$\int_2^6 \sin\left(\frac{3x}{2}\right)\delta(x - \pi)dx = f(\pi) = \sin\left(\frac{3\pi}{2}\right) = -1.$$

d)

$$\int_{-2}^2 \left(2x^3 + 1\right)\delta(4x)dx.$$

Since $0 \in (-2, 2)$ and $f(x) = 2x^3 + 1$, we have

$$\int_{-2}^2 \left(2x^3 + 1\right)\delta(4x)dx = \frac{1}{|4|}f(0) = \frac{1}{|4|}(2(0)^3 + 1) = \frac{1}{4}.$$

e)

$$\int_{-\infty}^{\infty} x^2\delta(2x + 1)dx.$$

This can be rewritten as

$$\int_{-\infty}^{\infty} x^2\delta(2x + 1)dx = \int_{-\infty}^{\infty} x^2\delta\left[2\left(x + \frac{1}{2}\right)\right]dx = \int_{-\infty}^{\infty} x^2\frac{1}{|2|}\delta\left(x + \frac{1}{2}\right)dx.$$

Since $-\frac{1}{2} \in (-\infty, \infty)$ and $f(x) = x^2$, we have

$$\int_{-\infty}^{\infty} x^2\frac{1}{|2|}\delta\left(x + \frac{1}{2}\right)dx = \frac{1}{|2|}f\left(-\frac{1}{2}\right) = \frac{1}{|2|}\left(-\frac{1}{2}\right)^2 = \frac{1}{8}.$$

f)

$$\int_0^a \delta(x - b)dx.$$

Here we have

$$\int_0^a \delta(x - b)dx = \begin{cases} 1 & \text{if } 0 < b < a \\ 0 & \text{otherwise} \end{cases}.$$

Problem 1.20. Suppose we have two vector fields $\vec{F}_1 = y^2\hat{z}$ and $\vec{F}_2 = x\hat{x} + y\hat{y} + z\hat{z}$. Calculate the divergence and curl of each. Which can be written as the gradient of a scalar and which can be written as the curl of a vector? Find a scalar and a vector potential.

Solution For $\vec{F}_1$, we have

$$\nabla \cdot \vec{F}_1 = \left(\frac{\partial}{\partial x}\hat{x} + \frac{\partial}{\partial y}\hat{y} + \frac{\partial}{\partial z}\hat{z} \right) \cdot (y^2\hat{z}) = \frac{\partial(y^2)}{\partial z} = 0$$

and

$$\nabla \times \vec{F}_1 = \begin{vmatrix} \hat{x} & \hat{y} & \hat{z} \\ \frac{\partial}{\partial x} & \frac{\partial}{\partial y} & \frac{\partial}{\partial z} \\ 0 & 0 & y^2 \end{vmatrix} = 2y\hat{x}.$$

For $\vec{F}_2$, we have

$$\nabla \cdot \vec{F}_2 = \left(\frac{\partial}{\partial x}\hat{x} + \frac{\partial}{\partial y}\hat{y} + \frac{\partial}{\partial z}\hat{z} \right) \cdot (x\hat{x} + y\hat{y} + z\hat{z}) = 1 + 1 + 1 = 3$$

and

$$\nabla \times \vec{F}_2 = \begin{vmatrix} \hat{x} & \hat{y} & \hat{z} \\ \frac{\partial}{\partial x} & \frac{\partial}{\partial y} & \frac{\partial}{\partial z} \\ x & y & z \end{vmatrix} = (0 - 0)\hat{x} + (0 - 0)\hat{y} + (0 - 0)\hat{z} = 0.$$

Since $\nabla \cdot \vec{F}_1 = 0$, $\vec{F}_1$ can be expressed as $\vec{F}_1 = \nabla \times \vec{A}$. We can find $\vec{A}$ by considering

$$\nabla \times \vec{A} = \begin{vmatrix} \hat{x} & \hat{y} & \hat{z} \\ \frac{\partial}{\partial x} & \frac{\partial}{\partial y} & \frac{\partial}{\partial z} \\ 0 & 0 & y^2 \end{vmatrix}$$

$$= \left(\frac{\partial A_z}{\partial y} - \frac{\partial A_y}{\partial z} \right)\hat{x} + \left(\frac{\partial A_x}{\partial z} - \frac{\partial A_z}{\partial x} \right)\hat{y} + \left(\frac{\partial A_y}{\partial x} - \frac{\partial A_x}{\partial y} \right)\hat{z}.$$

By inspection:

$$\frac{\partial A_z}{\partial y} - \frac{\partial A_y}{\partial z} = 0, \qquad \frac{\partial A_x}{\partial z} - \frac{\partial A_z}{\partial x} = 0, \qquad \frac{\partial A_y}{\partial x} - \frac{\partial A_x}{\partial y} = y^2.$$

This is satisfied by

$$\vec{A} = y^2 x\hat{y},$$

which is just one example. Since $\nabla \times \vec{F}_2 = 0$, $\vec{F}_2$ can be expressed as $\vec{F}_2 = -\nabla V$. We can find V by considering

$$\vec{F}_2 = -\left(\frac{\partial V}{\partial x}\hat{x} + \frac{\partial V}{\partial y}\hat{y} + \frac{\partial V}{\partial z}\hat{z}\right).$$

By inspection:

$$x = -\frac{\partial V}{\partial x}, \qquad y = -\frac{\partial V}{\partial y}, \qquad z = -\frac{\partial V}{\partial z}.$$

This is satisfied by

$$V = -\left(\frac{x^2}{2} + \frac{y^2}{2} + \frac{z^2}{2}\right)$$

which is again just one example.

Bibliography

Byron F W and Fuller R W 1992 *Mathematics of Classical and Quantum Physics* (New York: Dover)

Griffiths D J 1999 *Introduction to Electrodynamics* 3rd edn (Englewood Cliffs, NJ: Prentice Hall)

Griffiths D J 2013 *Introduction to Electrodynamics* 4th edn (New York: Pearson)

Halliday D, Resnick R and Walker J 2010 *Fundamentals of Physics* 9th edn (New York: Wiley)

Halliday D, Resnick R and Walker J 2013 *Fundamentals of Physics* 10th edn (New York: Wiley)

Purcell E M and Morin D J 2013 *Electricity and Magnetism* 3rd edn (Cambridge: Cambridge University Press)

Rogawski J 2011 *Calculus: Early Transcendentals* 2nd edn (San Francisco, CA: Freeman)

Chapter 2

Electrostatics

Electrostatics is the topic of this chapter. Coulomb's law, Gauss's law, and the energy of various charge distributions are a few ways of understanding the electric field. The methods employed will make use of the specific degrees of symmetry. The mathematical skills obtained in chapter 1 will be applied here to analyze different charge distributions in Cartesian, spherical, or cylindrical coordinates.

2.1 Theory

2.1.1 Coulomb's law

The force on a point charge q due to a charge Q, separated by a distance r, is given by

$$\vec{F} = \frac{1}{4\pi\varepsilon_o} q \frac{Q}{r^2} \hat{r},$$

where $\varepsilon_0 (= 8.85 \times 10^{-12} \frac{C^2}{Nm^2})$ is the permittivity of free space.

2.1.2 Electric field

In general, for a volume charge density $\rho(\vec{r})$, the electric field at $\vec{r}$ is given by

$$\vec{E}(\vec{r}) = \frac{1}{4\pi\varepsilon_o} \int_{\mathcal{V}} \frac{\rho(\vec{r}')}{r^2} \hat{r}\, d\tau'.$$

For a surface charge density $\sigma(\vec{r})$, the electric field is given by

$$\vec{E}(\vec{r}) = \frac{1}{4\pi\varepsilon_o} \int_{S} \frac{\sigma(\vec{r}')}{r^2} \hat{r}\, da'.$$

doi:10.1088/978-1-6817-4429-2ch2

For a linear charge density $\lambda(\vec{r})$, the electric field is given by

$$\vec{E}(\vec{r}) = \frac{1}{4\pi\varepsilon_o} \int_{\mathcal{P}} \frac{\lambda(\vec{r}')}{r^2} \hat{r} \, \mathrm{d}\ell'.$$

2.1.3 Gauss's law

For an electric field $\vec{E}$ and surface $\mathcal{S}$, Gauss's law states

$$\oint_{\mathcal{S}} \vec{E} \cdot \mathrm{d}\vec{a} = \frac{q_{enc}}{\varepsilon_0},$$

where the enclosed charge is

$$q_{enc} = \int_{\mathcal{V}} \rho \, \mathrm{d}\tau.$$

In differential form

$$\nabla \cdot \vec{E} = \frac{\rho}{\varepsilon_o},$$

where ρ is the volume charge density.

2.1.4 Curl of $\vec{E}$

$$\oint_{\mathcal{P}} \vec{E} \cdot \mathrm{d}\vec{\ell} = 0 \rightarrow \nabla \times \vec{E} = 0,$$

where $\vec{E}$ is an *electrostatic* field.

2.1.5 Energy of a point charge distribution

The energy required to assemble n charges $q_1, q_2, \ldots, q_n$ is given by

$$W = \frac{1}{2} \sum_{i=i}^{n} q_i \left(\sum_{\substack{j=i \\ j \neq i}}^{n} \frac{1}{4\pi\varepsilon_o} \frac{q_j}{r_{ij}} \right) = \frac{1}{2} \sum_{i=1}^{n} q_i V(\vec{r}_i),$$

where $V(\vec{r}_i)$ is the potential at charge q_i and r_{ij} is the distance between charges q_i and q_j.

2.1.6 Energy of a continuous distribution

$$W = \frac{1}{2} \int \rho V(\vec{r}) \, \mathrm{d}\tau = \frac{\varepsilon_o}{2} \int_{\substack{\text{all} \\ \text{space}}} E^2 \, \mathrm{d}\tau.$$

2.1.7 Energy per unit volume

$$\frac{W}{\text{volume}} = \frac{\varepsilon_o}{2} E^2$$

2.2 Problems and solutions

Problem 2.1. Given the charge distribution below, find the force on charge $q_1 = q$ with $q_2 = 3q$, $q_3 = -2q$, and $q_4 = q$.

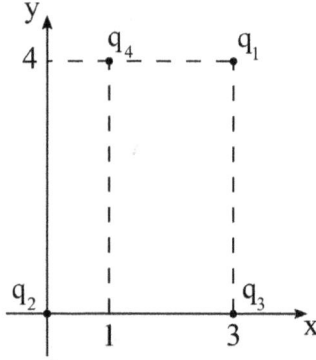

Solution The force on q_1 from q_2 is given by

$$\vec{F}_{21} = \frac{1}{4\pi\varepsilon_o} \frac{q_1 q_2}{r^2} \hat{r} = \frac{1}{4\pi\varepsilon_o} \frac{q_1 q_2}{r^3} \vec{r},$$

where

$$\vec{r} = 3\hat{x} + 4\hat{y}$$

and

$$r = \sqrt{3^2 + 4^2} = 5.$$

So

$$\vec{F}_{21} = \frac{1}{4\pi\varepsilon_o} \frac{3q^2}{5^3} (3\hat{x} + 4\hat{y}) = \frac{q^2}{4\pi\varepsilon_o} \left(\frac{9}{125}\hat{x} + \frac{12}{125}\hat{y} \right).$$

The force on q_1 from q_3 is given by

$$\vec{F}_{31} = \frac{1}{4\pi\varepsilon_o} \frac{q_1 q_3}{r^2} \hat{r},$$

where

$$r = 4$$

and

$$\hat{r} = \hat{y}.$$

So

$$\vec{F}_{31} = \frac{1}{4\pi\varepsilon_o}\left(-\frac{2q^2}{4^2}\right)\hat{y} = -\frac{q^2}{4\pi\varepsilon_o}\frac{1}{8}\hat{y}.$$

The force on q_1 from q_4 is given by

$$\vec{F}_{41} = \frac{1}{4\pi\varepsilon_o}\frac{q_1 q_4}{r^2}\hat{r},$$

where

$$r = 2$$

and

$$\hat{r} = \hat{x}.$$

So

$$\vec{F}_{41} = \frac{1}{4\pi\varepsilon_o}\frac{q^2}{2^2}\hat{r} = \frac{q^2}{4\pi\varepsilon_o}\frac{1}{4}\hat{x}.$$

Therefore,

$$\vec{F}_1 = \vec{F}_{21} + \vec{F}_{31} + \vec{F}_{41} = \frac{q^2}{4\pi\varepsilon_o}\left[\left(\frac{9}{125} + \frac{1}{4}\right)\hat{x} + \left(\frac{12}{125} - \frac{1}{8}\right)\hat{y}\right]$$

$$\vec{F}_1 = \frac{q^2}{4\pi\varepsilon_o}\left(\frac{161}{500}\hat{x} - \frac{29}{1000}\hat{y}\right).$$

Problem 2.2. Given a charged sheet with surface charge density $\sigma = ky$ (where k is a constant) and sides of length $2d$, find the electric field z above the center of the sheet.

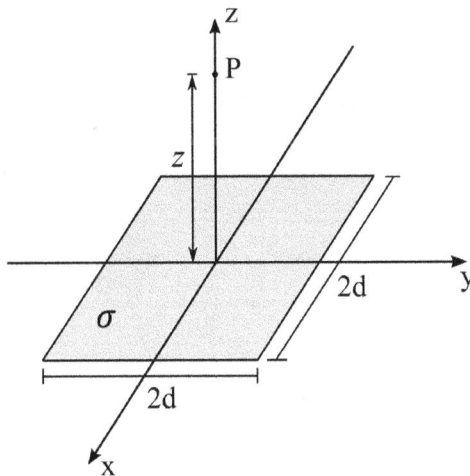

The electric field is given by

$$\vec{E} = \frac{1}{4\pi\varepsilon_o} \int \frac{\sigma}{r^2} \hat{r} \, da.$$

The horizontal components cancel so we only have the $\hat{z}$-component:

$$\hat{r} \rightarrow \cos\theta \, \hat{z} = \frac{z}{r}\hat{z}.$$

Also we have $da = dx \, dy$ and $r^2 = x^2 + y^2 + z^2$. Note that the piece of the sheet in each quadrant of the xy-plane contributes the same amount to the total field. Therefore,

$$\vec{E} = 4\vec{E}_{\text{quad}} = \frac{4}{4\pi\varepsilon_o} \int_0^d \int_0^d \frac{kyz\hat{z}}{\left(x^2 + y^2 + z^2\right)^{3/2}} dx \, dy = \frac{kz\hat{z}}{\pi\varepsilon_o} \int_0^d \int_0^d \frac{y}{(x^2 + y^2 + z^2)^{3/2}} dx \, dy$$

$$= \frac{kz\hat{z}}{\pi\varepsilon_o} \int_0^d y \left[\frac{x}{\left(y^2 + z^2\right)\sqrt{y^2 + z^2 + x^2}} \right]_{x=0}^{x=d} dy = \frac{kzd\hat{z}}{\pi\varepsilon_o} \int_0^d \frac{y}{\left(y^2 + z^2\right)\sqrt{y^2 + z^2 + d^2}} dy.$$

Let

$$u^2 = y^2 + z^2$$

so

$$2u \, du = 2y \, dy \rightarrow u \, du = y \, dy.$$

Evaluating u at the endpoints yields

$$u^2(y = 0) = z^2 \rightarrow u = z$$

$$u^2(y = d) = d^2 + z^2 \rightarrow u = \sqrt{d^2 + z^2}.$$

Now

$$\vec{E} = \frac{kzd\hat{z}}{\pi\varepsilon_o} \int_z^{\sqrt{z^2+d^2}} \frac{du}{u\sqrt{u^2 + d^2}}$$

$$= \frac{kzd\hat{z}}{\pi\varepsilon_o} \left[\frac{1}{d} \ln\left(\frac{d + \sqrt{d^2 + u^2}}{u} \right) \right]_{u=\sqrt{z^2+d^2}}^{u=z} = \frac{kz\hat{z}}{\pi\varepsilon_o} \ln\left(\frac{\dfrac{d + \sqrt{d^2 + z^2}}{z}}{\dfrac{d + \sqrt{2d^2 + z^2}}{\sqrt{z^2 + d^2}}} \right)$$

$$\vec{E} = \frac{kz}{\pi\varepsilon_o} \ln\left[\frac{\left(d + \sqrt{z^2 + d^2}\right)\sqrt{z^2 + d^2}}{z\left(d + \sqrt{2d^2 + z^2}\right)} \right]\hat{z}.$$

Problem 2.3. Find the electric field d above a cylinder of radius R, height h, and volume density ρ (ignoring edge effects).

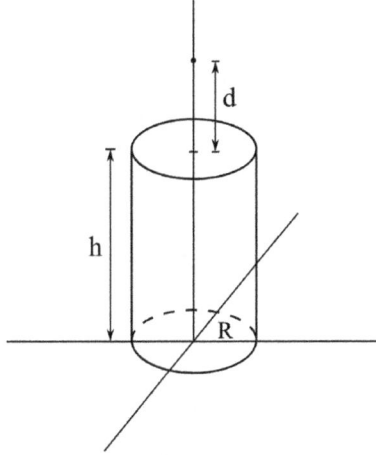

Solution We have

$$\vec{E} = \frac{1}{4\pi\varepsilon_o} \int \frac{\rho}{r^2} \hat{r} \, d\tau.$$

Note our horizontal components cancel, so $\hat{r} \rightarrow \cos\theta \, \hat{z}$ with

$$\cos\theta = \frac{d+h-z}{r}.$$

Also

$$d\tau = s \, ds \, d\phi \, dz$$

and

$$r^2 = s^2 + (d+h-z)^2.$$

Therefore,

$$\vec{E} = \frac{\rho}{4\pi\varepsilon_o} \int_0^{2\pi} \int_0^R \int_0^h \frac{(d+h-z)s \, \hat{z}}{\left[s^2 + (d+h-z)^2\right]^{3/2}} dz \, ds \, d\phi$$

$$= \frac{2\pi\rho}{4\pi\varepsilon_o}\left[\sqrt{R^2 + d^2} - \sqrt{R^2 + (d+h)^2} + h\right]\hat{z}$$

$$\vec{E} = \frac{\rho}{2\varepsilon_o}\left[\sqrt{R^2 + d^2} - \sqrt{R^2 + (d+h)^2} + h\right]\hat{z}.$$

Note if $R \gg d$ and $R \gg h$, the field reduces to

$$\vec{E} = \frac{\rho h}{2\varepsilon_o}\hat{z},$$

which is the field given by an infinite sheet of surface charge $\sigma = \rho h$.

Problem 2.4. Given the bottom hemisphere of a spherical shell of radius R, thickness d, and volume charge density ρ, find the electric field z above the center (above the open part, ignoring edge effects).

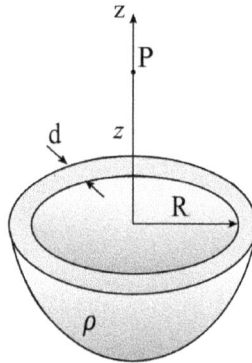

Solution The electric field is given by

$$\vec{E} = \frac{1}{4\pi\varepsilon_o} \int \frac{\rho}{r^2}\hat{r}\, d\tau.$$

We can see from

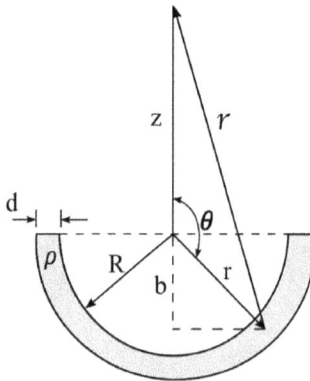

that

$$R \leqslant r \leqslant R + d, \quad 0 \leqslant \phi \leqslant 2\pi, \quad \frac{\pi}{2} \leqslant \theta \leqslant \pi.$$

Also, $d\tau = r^2 \sin\theta \, dr \, d\phi \, d\theta$. From the law of cosines,

$$r^2 = z^2 + r^2 - 2rz \cos\theta.$$

Since the horizontal components cancel, $\hat{r}$ becomes

$$\hat{r} \rightarrow \cos\gamma \, \hat{z} = \frac{z+b}{r}\hat{z},$$

where b is given by

$$\cos(\pi - \theta) = \frac{b}{r} \rightarrow b = -r\cos\theta.$$

So

$$\hat{r} \rightarrow \frac{z - r\cos\theta}{r}\hat{z}.$$

Therefore,

$$\vec{E} = \frac{\rho}{4\pi\varepsilon_o} \int_R^{R+d} \int_{\frac{\pi}{2}}^{\pi} \int_0^{2\pi} \frac{r^2(z - r\cos\theta)\sin\theta \, \hat{z}}{\left(z^2 + r^2 - 2rz\cos\theta\right)^{3/2}} d\phi \, d\theta \, dr$$

$$= \frac{\rho\hat{z}}{2\varepsilon_o} \int_R^{R+d} \int_{\frac{\pi}{2}}^{\pi} \frac{r^2(z - r\cos\theta)\sin\theta \, \hat{z}}{\left(z^2 + r^2 - 2rz\cos\theta\right)^{3/2}} d\theta \, dr.$$

Let $u = \cos\theta$ and $du = -\sin\theta \, d\theta$

$$u\left(\theta = \frac{\pi}{2}\right) = 0$$

$$u(\theta = \pi) = -1$$

$$\vec{E} = \frac{\rho\hat{z}}{2\varepsilon_o} \int_R^{R+d} \int_{-1}^0 \frac{r^2(z - ru)\hat{z}}{\left(z^2 + r^2 - 2rzu\right)^{3/2}} du \, dr$$

$$= \frac{\rho\hat{z}}{2\varepsilon_o z^2} \int_R^{R+d} \left(r^2 - \frac{r^3}{\sqrt{r^2 + z^2}}\right) dr$$

$$= \frac{\rho\hat{z}}{2\varepsilon_o z^2}\left[\frac{r^3}{3} - \frac{\sqrt{r^2 + z^2}\left(r^2 - 2z^2\right)}{3}\right]_{r=R}^{r=R+d}$$

$$\vec{E} = \frac{\rho\hat{z}}{6\varepsilon_o z^2}\left\{(R+d)^3 - \sqrt{(R+d)^2 + z^2}\left[(R+d)^2 - 2z^2\right]\right.$$

$$\left. - R^3 + \sqrt{\left(R^2 + z^2\right)}\left(R^2 - 2z^2\right)\right\}.$$

Problem 2.5. Given the electric field $\vec{E} = k[2xz\hat{x} + z^2\hat{y} + (x^2 + 2yz)\hat{z}]$ (with constant k) find the following:
a) The charge density ρ.
b) The charge enclosed by a cylinder of height h, radius R, and base on the xy-plane center at the origin (below).
c) The charge enclosed by an upper hemisphere of radius R centered at the origin.

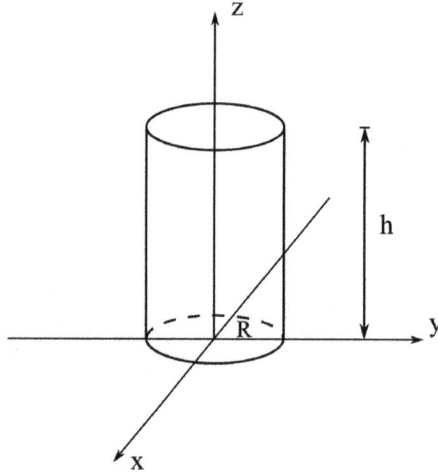

Solutions
a) The charge density ρ.
      Gauss's law states

$$\nabla \cdot \vec{E} = \frac{\rho}{\varepsilon_o}.$$

So

$$\rho = \varepsilon_o \, \nabla \cdot \vec{E} = k\varepsilon_o(2y + 2z) = 2k\varepsilon_o(y + z).$$

b) The charge enclosed by a cylinder of height h, radius R, and base on the xy-plane center at the origin.
      We have

$$q_{enc} = \int_v \rho \, d\tau,$$

with

$$\rho = 2k\varepsilon_o(y + z).$$

We can transform ρ into cylindrical coordinates using $x = s \cos \phi$, $y = s \sin \phi$, and $z = z$.

So

$$q_{enc} = 2k\varepsilon_o \int_0^R \int_0^h \int_0^{2\pi} (s \sin \phi + z)s \, d\phi \, dz \, ds = \pi k\varepsilon_o h^2 R^2.$$

c) The charge enclosed by an upper hemisphere of radius R centered at the origin.

Again

$$q_{enc} = \int_V \rho \, d\tau$$

with

$$\rho = 2k\varepsilon_o(y + z),$$

but now $y = r \sin \phi \sin \theta$ and $z = r \cos \theta$. So

$$q_{enc} = 2k\varepsilon_o \int_0^R \int_0^{2\pi} \int_0^{\frac{\pi}{2}} r(\sin \phi \sin \theta + \cos \theta)r^2 \sin \theta \, d\theta \, d\phi \, dr = \frac{k\varepsilon_o \pi R^4}{2}.$$

Problem 2.6. Given a charge q located in the center of a spherical shell of radius R and surface charge $\sigma = k \sin \theta$ (with constant k), find the electric field inside and outside the shell.

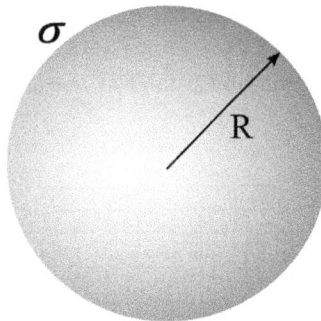

Solution We will use Gauss's law to find the field

$$\oint_S \bar{E} \cdot d\bar{a} = \frac{q_{enc}}{\varepsilon_o},$$

where

$$\oint_S \vec{E} \cdot \mathrm{d}\vec{a} = \oint_S E \, \mathrm{d}a = E \oint_S \mathrm{d}a = E 4\pi r^2.$$

For $r < R$, we have

$$q_{\mathrm{enc}} = q.$$

So

$$E 4\pi r^2 = \frac{q}{\varepsilon_o} \rightarrow \vec{E} = \frac{1}{4\pi\varepsilon_o}\frac{q}{r^2}\hat{r}.$$

For $r > R$, we have

$$q_{\mathrm{enc}} = q + \int \sigma \, \mathrm{d}a = q + k \int_0^{2\pi} \int_0^{\pi} R^2 \sin^2\theta \, \mathrm{d}\theta \, \mathrm{d}\phi.$$

$$q_{\mathrm{enc}} = q + k R^2 \pi^2.$$

So

$$E 4\pi r^2 = \frac{q}{\varepsilon_o} + \frac{k R^2 \pi^2}{\varepsilon_o}$$

and

$$\vec{E} = \frac{1}{4\pi\varepsilon_o}\frac{q + k R^2 \pi^2}{r^2}\hat{r}.$$

Problem 2.7. Given a line of charge carrying λ surrounded by a cylindrical shell with inner radius a, outer radius b, and charge density $\rho = ks^2$, find the electric field in the regions $s < a$, $a < s < b$, and $b < s$.

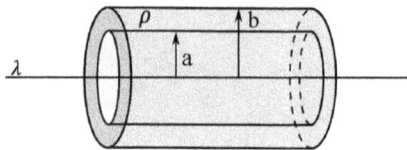

Solution Gauss's law states

$$\oint_S \vec{E} \cdot \mathrm{d}\vec{a} = \frac{q_{\mathrm{enc}}}{\varepsilon_o},$$

where

$$\oint_S \vec{E} \cdot \mathrm{d}\vec{a} = \oint_S E \, \mathrm{d}a = E \oint_S \mathrm{d}a = E2\pi s\ell.$$

For $s < a$, we have

$$q_{\mathrm{enc}} = \lambda\ell.$$

So

$$E2\pi s\ell = \frac{\lambda\ell}{\varepsilon_o}$$

and

$$\vec{E} = \frac{\lambda}{2\pi\varepsilon_o s}\hat{s}.$$

For $a < s < b$, we have

$$q_{\mathrm{enc}} = \lambda\ell + \int \rho \, \mathrm{d}\tau = \lambda\ell + \int_0^\ell \int_0^{2\pi} \int_a^s k(s')^2 s' \, \mathrm{d}s' \, \mathrm{d}\phi' \, \mathrm{d}z' = \ell\left[\lambda + \frac{k\pi}{2}\left(s^4 - a^4\right)\right].$$

So

$$\oint_S \vec{E} \cdot \mathrm{d}\vec{a} = \frac{q_{\mathrm{enc}}}{\varepsilon_o} \rightarrow E2\pi s\ell = \frac{\ell\left[\lambda + \frac{k\pi}{2}\left(s^4 - a^4\right)\right]}{\varepsilon_o}$$

and

$$\vec{E} = \frac{2\lambda + k\pi\left(s^4 - a^4\right)}{4\pi\varepsilon_o s}\hat{s}.$$

For $b < s$, we have

$$q_{\mathrm{enc}} = \lambda\ell + \int \rho \, \mathrm{d}\tau = \lambda\ell + \int_0^\ell \int_0^{2\pi} \int_a^b k(s')^2 s' \, \mathrm{d}s' \, \mathrm{d}\phi' \, \mathrm{d}z' = \ell\left[\lambda + \frac{k\pi}{2}\left(b^4 - a^4\right)\right].$$

So

$$\oint_S \vec{E} \cdot \mathrm{d}\vec{a} = \frac{q_{\mathrm{enc}}}{\varepsilon_o} \rightarrow E2\pi s\ell = \frac{\ell\left[\lambda + \frac{k\pi}{2}\left(b^4 - a^4\right)\right]}{\varepsilon_o}$$

and

$$\vec{E} = \frac{2\lambda + k\pi\left(b^4 - a^4\right)}{4\pi\varepsilon_o s}\hat{s}.$$

Problem 2.8. Which of the following is a possible electrostatic field?
a) $\vec{E} = k(yz\hat{x} + xz\hat{y} + x^2\hat{z})$
b) $\vec{E} = k(x\hat{x} + y\hat{y} + z\hat{z})$
c) $\vec{E} = k[2xz\hat{x} + z^2\hat{y} + (x^2 + 2yz)\hat{z}]$

where k is a constant with the appropriate units for the given field. For the possible electric field, find the electric potential using the origin as your reference point. Check your answer by verifying that $\vec{E} = -\nabla V$.

Solutions

a) $\vec{E} = k\left(yz\hat{x} + xz\hat{y} + x^2\hat{z}\right)$

$$\nabla \times \vec{E} = \begin{vmatrix} \hat{x} & \hat{y} & \hat{z} \\ \dfrac{\partial}{\partial x} & \dfrac{\partial}{\partial y} & \dfrac{\partial}{\partial z} \\ yz & xz & x^2 \end{vmatrix}$$

$$= \left[\frac{\partial}{\partial y}(x^2) - \frac{\partial}{\partial z}(xz)\right]\hat{x} + \left[\frac{\partial}{\partial z}(yz) - \frac{\partial}{\partial x}(x^2)\right]\hat{y} + \left[\frac{\partial}{\partial x}(xz) - \frac{\partial}{\partial y}(yz)\right]\hat{z}$$

$$= (0 - x)\hat{x} + (y - 2x)\hat{y} + (z - z)\hat{z} = -x\hat{x} + (-2x + y)\hat{y}.$$

Since $\nabla \times \vec{E} \neq 0$, this is *not* a possible electric field.

b) $\vec{E} = k(x\hat{x} + y\hat{y} + z\hat{z})$

$$\nabla \times \vec{E} = \begin{vmatrix} \hat{x} & \hat{y} & \hat{z} \\ \dfrac{\partial}{\partial x} & \dfrac{\partial}{\partial y} & \dfrac{\partial}{\partial z} \\ x & y & z \end{vmatrix}$$

$$= \left[\frac{\partial}{\partial y}(z) - \frac{\partial}{\partial z}(y)\right]\hat{x} + \left[\frac{\partial}{\partial z}(x) - \frac{\partial}{\partial x}(z)\right]\hat{y} + \left[\frac{\partial}{\partial x}(y) - \frac{\partial}{\partial y}(x)\right]\hat{z} = 0.$$

Since $\nabla \times \vec{E} = 0$, this is a possible electric field. Let us find the electric potential by integrating along the path given by

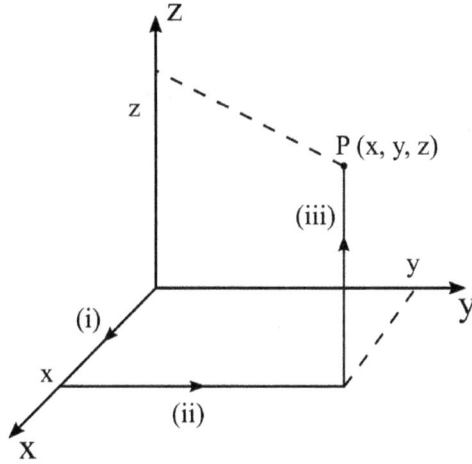

The potential is given by

$$V = -\int_{\mathcal{O}}^{\vec{r}} \vec{E} \cdot d\vec{l}.$$

where

$$\vec{E} \cdot d\vec{l} = k(x\hat{x} + y\hat{y} + z\hat{z}) \cdot (dx\hat{x} + dy\hat{y} + dz\hat{z})$$
$$= k(x\,dx + y\,dy + z\,dz).$$

Note along (i) we only have dx, along (ii) we only have dy, and along (iii) we only have dz. Therefore, taking the origin $\mathcal{O} = (0, 0, 0)$ as our reference point, the potential will be given by

$$V = -\int_{\mathcal{O}}^{\vec{r}} \vec{E} \cdot d\vec{l} = -k\left(\underbrace{\int_0^x x'dx'}_{(i)} + \underbrace{\int_0^y y'dy'}_{(ii)} + \underbrace{\int_0^z z'dz'}_{(iii)}\right).$$

So,

$$V(r) = -\frac{k}{2}\left(x^2 + y^2 + z^2\right).$$

We can check this using

$$\vec{E} = - \nabla V$$

$$= -\left\{ \frac{\partial}{\partial x}\left[-\frac{k}{2}(x^2 + y^2 + z^2) \right]\hat{x} + \frac{\partial}{\partial y}\left[-\frac{k}{2}(x^2 + y^2 + z^2) \right]\hat{y} \right.$$

$$\left. + \frac{\partial}{\partial z}\left[-\frac{k}{2}(x^2 + y^2 + z^2) \right]\hat{z} \right\}$$

$$= \frac{2kx}{2}\hat{x} + \frac{2ky}{2}\hat{y} + \frac{2kz}{2}\hat{z}$$

$$= k(x\hat{x} + y\hat{y} + z\hat{z}) = \vec{E}.$$

c) $\vec{E} = k[2xz\hat{x} + z^2\hat{y} + (x^2 + 2yz)\hat{z}]$

$$\nabla \times \vec{E} = \begin{vmatrix} \hat{x} & \hat{y} & \hat{z} \\ \dfrac{\partial}{\partial x} & \dfrac{\partial}{\partial y} & \dfrac{\partial}{\partial z} \\ 2xz & z^2 & x^2 + 2yz \end{vmatrix}$$

$$= \left[\frac{\partial}{\partial y}(x^2 + 2yz) - \frac{\partial}{\partial z}(z^2) \right]\hat{x} + \left[\frac{\partial}{\partial z}(2xz) - \frac{\partial}{\partial x}(x^2 + 2yz) \right]\hat{y}$$

$$+ \left[\frac{\partial}{\partial x}(z^2) - \frac{\partial}{\partial y}(2xz) \right]\hat{z}$$

$$= k\left[(2z - 2z)\hat{x} + (2x - 2x)\hat{y} + (0 - 0)\hat{z} \right] = 0.$$

Since $\nabla \times \vec{E} = 0$, this is a possible electric field. Let us find the electric potential by integrating along the same path as before. The potential is given by

$$V = -\int_{\mathcal{O}}^{\vec{r}} \vec{E} \cdot d\vec{\ell}$$

where

$$\vec{E} \cdot d\vec{\ell} = k\left[2xz\hat{x} + z^2\hat{y} + \left(x^2 + 2yz \right)\hat{z} \right] \cdot (dx\hat{x} + dy\hat{y} + dz\hat{z})$$

$$= k\left[2xzdx + z^2dy + \left(x^2 + 2yz \right)dz \right].$$

Note along (i) we only have dx with $y = 0$ and $z = 0$, along (ii) we only have dy with $x = 1$ and $z = 0$, and along (iii) we only have dz with $x = 1$ and $y = 1$. Therefore, taking the origin $\mathcal{O} = (0, 0, 0)$ as our reference point, the potential will be given by

$$V = - \int_{\mathcal{O}}^{\vec{r}} \vec{E} \cdot d\vec{\ell} = -k \left(\underbrace{\int_0^x 2x'z\ dx'}_{(i)} + \underbrace{\int_0^y z^2 dy'}_{(ii)} + \underbrace{\int_0^z (2yz' + x^2)dz'}_{(iii)} \right)$$

$$= -k \left(\underbrace{\int_0^x 2x'(0)dx'}_{(i)} + \underbrace{\int_0^y 0^2\ dy'}_{(ii)} + \underbrace{\int_0^z (2(1)z'+1^2)dz'}_{(iii)} \right)$$

$$V(r) = -k\left(yz^2 + x^2z \right).$$

We can check this using

$$\vec{E} = -\nabla V$$

$$= - \left\{ \frac{\partial}{\partial x}\left[-k\left(yz^2 + x^2z \right) \right]\hat{x} + \frac{\partial}{\partial y}\left[-k\left(yz^2 + x^2z \right) \right]\hat{y} + \frac{\partial}{\partial z}\left[-k\left(yz^2 + x^2z \right) \right]\hat{z} \right\}$$

$$= k\left[(2xz)\hat{x} + \left(z^2 \right)\hat{y} + \left(x^2 + 2yz \right)\hat{z} \right] = \vec{E}.$$

Problem 2.9. Find the electric field and the electric potential inside and outside a thin spherical shell of radius R that carries a uniform surface charge σ. Set the reference point at infinity.

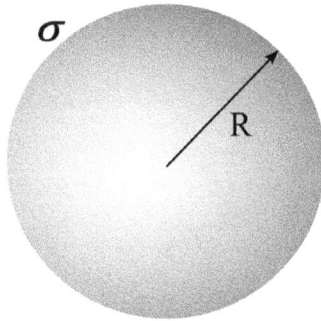

Solution Let us find the electric field everywhere by using Gauss's law, given by

$$\oint_S \vec{E} \cdot d\vec{a} = \frac{q_{enc}}{\varepsilon_0},$$

where

$$\oint_S \vec{E} \cdot d\vec{a} = \oint_S E \, da = E \oint_S da = E 4 \pi r^2.$$

For $r < R$, we have our Gaussian surface given by

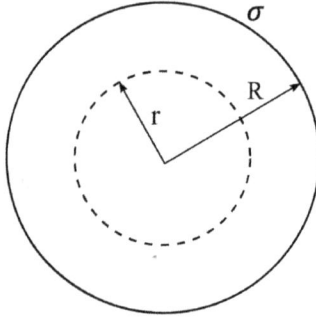

where r is the radius of the Gaussian sphere with radius smaller than R. Note that

$$q_{\text{enc}} = 0.$$

So,

$$E 4 \pi r^2 = 0 \rightarrow \vec{E} = 0.$$

For $r > R$, we have our Gaussian surface given by

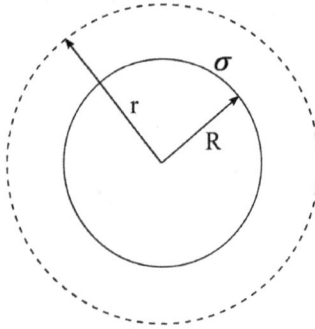

Now we have

$$q_{\text{enc}} = \sigma 4 \pi R^2.$$

So,

$$E 4 \pi r^2 = \frac{\sigma 4 \pi R^2}{\varepsilon_0}$$

and

$$\vec{E} = \frac{\sigma R^2}{\varepsilon_0 r^2}\hat{r}.$$

Now let us calculate the electric potential everywhere taking the reference point at infinity. We will use

$$V = -\int_{\infty}^{r} \vec{E} \cdot d\vec{\ell},$$

where

$$d\vec{\ell} = dr\,\hat{r} + r\,d\theta\,\hat{\theta} + r\sin\theta\,d\phi\,\hat{\phi}.$$

For $r > R$,

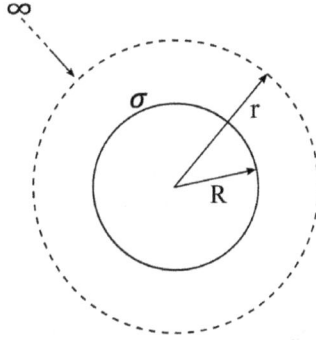

therefore,

$$V = -\int_{\infty}^{r} \vec{E} \cdot d\vec{\ell} = -\int_{\infty}^{r} \frac{\sigma R^2}{\varepsilon_0 r^2}\hat{r} \cdot \left(dr\,\hat{r} + r\,d\theta\,\hat{\theta} + r\sin\theta\,d\phi\,\hat{\phi}\right) = -\int_{\infty}^{r} \frac{\sigma R^2}{\varepsilon_0 r'^2}dr'$$

$$V = \frac{\sigma R^2}{\varepsilon_0 r}.$$

For $r < R$

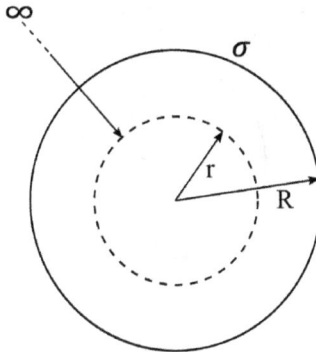

$$V = -\int_{\infty}^{\vec{r}} \vec{E} \cdot d\vec{\ell} = -\int_{\infty}^{R} \frac{\sigma R^2}{\varepsilon_0 r} dr - \int_{R}^{r} 0 \, dr' = \frac{\sigma R^2}{\varepsilon_0 R} - 0 = \frac{\sigma R}{\varepsilon_0} = \text{const.}$$

Note that the potential inside the shell is constant, as the electric field is zero.

Problem 2.10. Calculate the electric field and the electric potential inside and outside a solid sphere of radius R having a uniform charge distribution ρ. Use infinity as your reference point. Then obtain the gradient of the potential everywhere and check that $\vec{E} = -\nabla V$. Plot the potential versus distance from the center of the sphere.

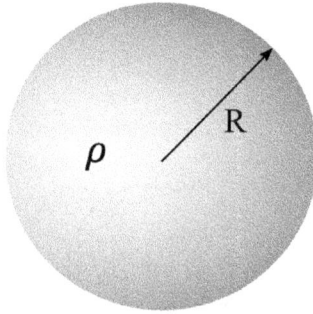

Solution Starting with the electric field, we use Gauss's law, given by

$$\oint_S \vec{E} \cdot d\vec{a} = \frac{q_{\text{enc}}}{\varepsilon_0},$$

where

$$\oint_S \vec{E} \cdot d\vec{a} = \oint_S E \, da = E \oint_S da = E 4\pi r^2.$$

For $r > R$, we have our Gaussian surface given by

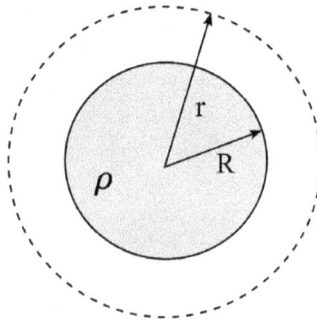

Here we simply have

$$q_{\text{enc}} = \rho \mathcal{V}_{\text{sp}} = \frac{\rho 4\pi R^3}{3},$$

where $\mathcal{V}_{\text{sp}}$ is the volume of the sphere. So

$$E 4\pi r^2 = \frac{\rho 4\pi R^3}{3\varepsilon_0}$$

and

$$\vec{E} = \frac{\rho R^3}{3\varepsilon_0 r^2}\hat{r}.$$

For $r < R$, our Gaussian surface becomes

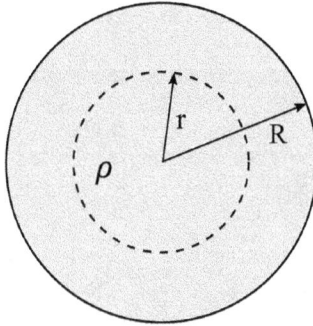

Now,

$$q_{\text{enc}} = \rho \mathcal{V}_{\text{enc}} = \frac{\rho 4\pi r^3}{3},$$

where $\mathcal{V}_{\text{enc}}$ is the enclosed volume. So

$$E 4\pi r^2 = \frac{\rho 4\pi r^3}{3\varepsilon_0}$$

and

$$\vec{E} = \frac{\rho r}{3\varepsilon_0}\hat{r}.$$

The plot of electric field is given by

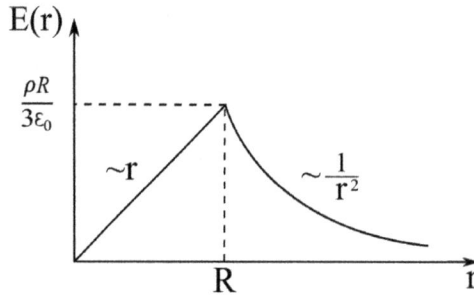

Now we can calculate the electric potential. This is done using

$$V = -\int_{\infty}^{\vec{r}} \vec{E} \cdot d\vec{\ell}$$

with infinity as the reference point. For $r > R$

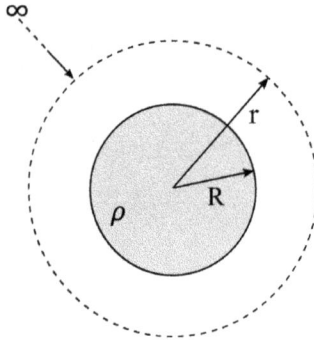

with

$$V = -\int_{\infty}^{\vec{r}} \vec{E} \cdot d\vec{\ell} = -\int_{\infty}^{r} \frac{\rho R^3}{3\varepsilon_0 r'^2} dr' = \frac{\rho R^3}{3\varepsilon_0 r}.$$

For $r < R$

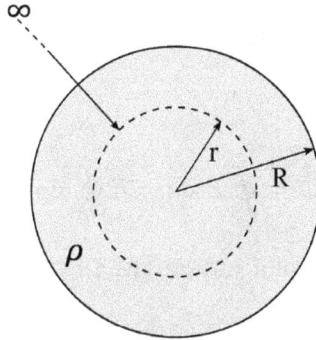

with

$$V = -\int_{\infty}^{\vec{r}} \vec{E} \cdot d\vec{\ell} = -\int_{\infty}^{R} \frac{\rho R^2}{3\varepsilon_0 r^2} dr - \int_{R}^{r} \frac{\rho r'}{3\varepsilon_0} dr' = \frac{\rho R^2}{2\varepsilon_0} - \frac{\rho r^2}{6\varepsilon_0}.$$

We can check using $\vec{E} = -\nabla V$. For $r > R$,

$$\vec{E} = -\nabla V = -\frac{\partial}{\partial r}\left(\frac{\rho R^2}{2\varepsilon_0} - \frac{\rho r^2}{6\varepsilon_0}\right)\hat{r} = \frac{\rho r}{3\varepsilon_0}\hat{r}$$

and for $r < R$,

$$\vec{E} = -\nabla V = -\frac{\partial}{\partial r}\left(\frac{\rho R^3}{2\varepsilon_0 r}\right)\hat{r} = \frac{\rho R^3}{3\varepsilon_0 r^2}\hat{r}$$

both of which are in agreement with what we found from Gauss's law.

Problem 2.11. Calculate the electric field and the electric potential for a sphere of radius R that carries a charge density $\rho = kr^2$, where k is a constant.

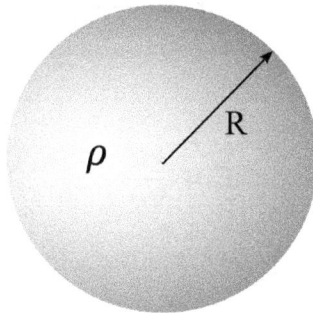

Solution Starting with the electric field, we use Gauss's law, given by

$$\oint_S \vec{E} \cdot d\vec{a} = \frac{q_{\text{enc}}}{\varepsilon_0},$$

where

$$\oint_S \vec{E} \cdot d\vec{a} = \oint_S E \, da = E \oint_S da = E 4\pi r^2.$$

For $r > R$, we have our Gaussian surface given by

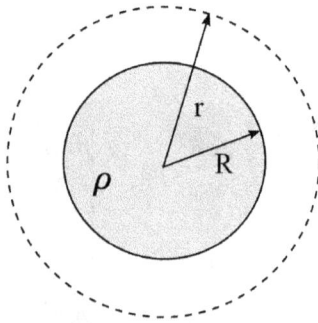

and our enclosed charge is given by

$$q_{\text{enc}} = \int \rho \, d\tau = \int_0^{2\pi} d\phi \int_0^{\pi} \sin\theta \, d\theta \int_0^R kr^2 r^2 dr = 4\pi k \frac{R^5}{5}.$$

Therefore,

$$E 4\pi r^2 = \frac{4\pi k R^5}{5\varepsilon_0}$$

so

$$\vec{E} = \frac{kR^5}{5\varepsilon_0 r^2}\hat{r}.$$

For $r < R$, our Gaussian surface becomes

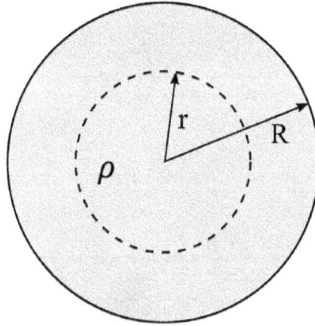

Now,

$$q_{\text{enc}} = \int \rho \, d\tau = \int\limits_{0}^{2\pi} d\phi \int\limits_{0}^{\pi} \sin\theta \, d\theta \int\limits_{0}^{R} kr'^2 r'^2 dr' = \frac{4\pi kr^5}{5}$$

so

$$E4\pi r^2 = \frac{4\pi kr^5}{5\varepsilon_0}$$

and

$$\vec{E} = \frac{kr^3}{5\varepsilon_0}\hat{r}.$$

Now we can calculate the electric potential. This is done using

$$V = -\int\limits_{\infty}^{\vec{r}} \vec{E} \cdot d\vec{\ell}$$

with infinity as the reference point. For $r > R$

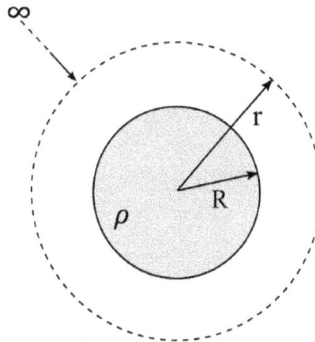

with

$$V = -\int_{\infty}^{\vec{r}} \vec{E} \cdot d\vec{\ell} = -\int_{\infty}^{r} \frac{kR^5}{5\varepsilon_0 r'^2} dr' = \frac{kR^5}{5\varepsilon_0 r}.$$

For $r < R$

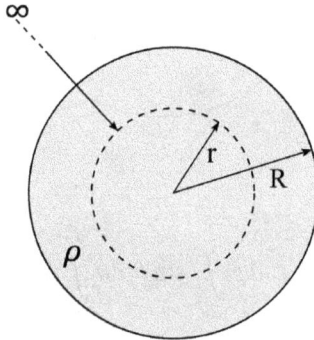

with

$$V = -\int_{\infty}^{\vec{r}} \vec{E} \cdot d\vec{\ell} = -\int_{\infty}^{R} \frac{kR^5}{5\varepsilon_0 r'^2} dr' - \int_{R}^{r} \frac{kr'^3}{5\varepsilon_0} dr' = \frac{kR^5}{5\varepsilon_0 R} - 0 - \frac{kr^4}{20\varepsilon_0} + \frac{kR^4}{20\varepsilon_0}$$

$$= \frac{k}{20\varepsilon_0}\left(4R^4 - r^4 + R^4\right)$$

$$V = \frac{kR^4}{20\varepsilon_0}\left(5 - \frac{r^4}{R^4}\right).$$

Problem 2.12 A long cylinder of radius a carries a charge density $\rho = ks^2$, where k is a constant and s is the distance from the axis of the cylinder. Find the electric field and the electric potential everywhere. Take the reference point at a distance b from the axis ($b > a$).

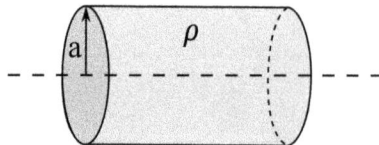

Solution Starting with the electric field, we use Gauss's law, given by

$$\oint_S \vec{E} \cdot d\vec{a} = \frac{q_{\text{enc}}}{\varepsilon_0}.$$

Note the left-hand side is always given by

$$\oint_S \vec{E} \cdot d\vec{a} = \oint_S E \, da = E \oint_S da = E2\pi s\ell.$$

For $s > a$, we have our Gaussian surface given by

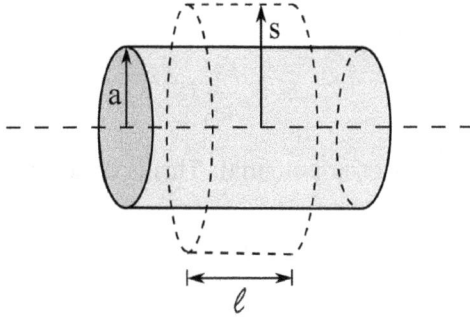

with enclosed charge given by

$$q_{\text{enc}} = \int \rho \, d\tau = \int_0^{2\pi} d\phi \int_0^{\ell} dz \int_0^a ks^2 s \, ds = \frac{\pi k \ell a^4}{2}.$$

Therefore,

$$E2\pi s\ell = \frac{\pi k \ell a^4}{2\varepsilon_0}$$

and

$$\vec{E} = \frac{ka^4}{4\varepsilon_0 s}\hat{s}.$$

For $s < a$, our Gaussian surface becomes

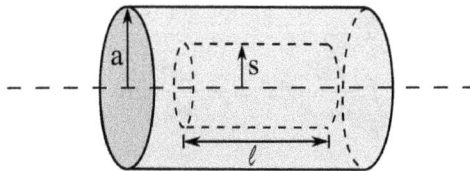

Now,

$$q_{\text{enc}} = \int \rho \, d\tau = \int\limits_0^{2\pi} d\phi \int\limits_0^{\ell} dz \int\limits_0^s ks'^2 s' \, ds' = \frac{\pi k \ell s^4}{2}$$

so

$$E 2\pi s \ell = \frac{\pi k \ell s^4}{2\varepsilon_0}$$

and

$$\vec{E} = \frac{ks^3}{4\varepsilon_0}\hat{s}.$$

Now we can calculate the electric potential. This is done using

$$V = -\int\limits_{\vec{b}}^{\vec{r}} \vec{E} \cdot d\vec{\ell}$$

with b as the reference point. For $s > a$,

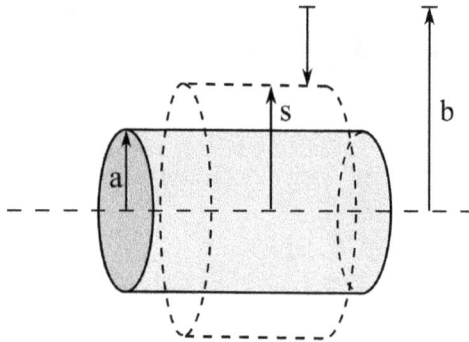

with

$$V = -\int\limits_{\vec{b}}^{\vec{r}} \vec{E} \cdot d\vec{\ell} = -\int\limits_b^s \frac{ka^4}{4\varepsilon_0 s'} ds' = -\frac{ka^4}{4\varepsilon_0}(\ln s - \ln b) = -\frac{ka^4}{4\varepsilon_0}\ln\frac{s}{b}.$$

For $s < a$,

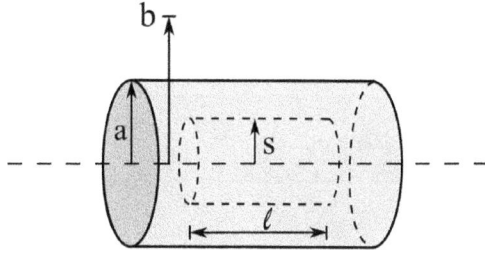

with

$$V = -\int \vec{E} \cdot d\vec{\ell} = -\int_b^a \frac{ka^4}{4\varepsilon_0 s}ds - \int_a^s \frac{ks'^3}{4\varepsilon_0}ds' = -\frac{ka^4}{4\varepsilon_0}\ln\frac{a}{b} - \frac{k(s^4 - a^4)}{16\varepsilon_0}$$

$$V = \frac{ka^4}{4\varepsilon_0}\ln\frac{b}{a} + \frac{k\left(a^4 - s^4\right)}{16\varepsilon_0}.$$

Problem 2.13. Verify the electrostatic boundary condition using the charge distribution in problem 2.9.

Solution The electrostatic boundary condition is given by

$$\vec{E}_{\text{above}} - \vec{E}_{\text{below}} = \frac{\sigma}{\varepsilon_0}\hat{n}.$$

From problem 2.9, our electric fields are

$$\vec{E} = \begin{cases} 0 & r < R \\ \dfrac{\sigma R^2}{\varepsilon_0 r^2}\hat{r} & r > R \end{cases}.$$

At $r = R$, we have

$$\vec{E}_{\text{above}} = \frac{\sigma}{\varepsilon_0}\hat{r}$$

and

$$\vec{E}_{\text{below}} = 0.$$

Therefore,

$$\vec{E}_{\text{above}} - \vec{E}_{\text{below}} = \frac{\sigma}{\varepsilon_0}\hat{r} - 0 = \frac{\sigma}{\varepsilon_0}\hat{n},$$

where $\hat{n}$ is normal to the sphere which has the direction of $\hat{r}$.

Problem 2.14. Find the work required to assemble the charge distribution below.

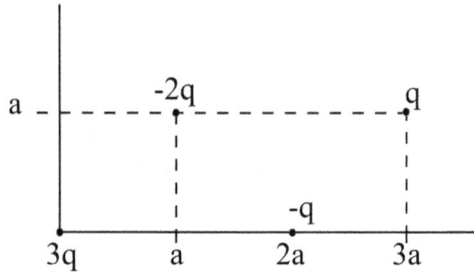

Solution We can denote the following $q_1 = 3q$, $q_2 = -2q$, $q_3 = -q$, $q_4 = q$. Starting with q_1, $W = 0$. Moving in q_2, we have

$$W_2 = \frac{1}{4\pi\varepsilon_0}q_2\left(\frac{q_1}{r_{12}}\right) = -\frac{6q^2}{4\pi\varepsilon_0}\frac{1}{a\sqrt{2}} = -\frac{q^2}{4\pi\varepsilon_0}\frac{3\sqrt{2}}{a}.$$

Moving in q_3, we have

$$W_3 = \frac{1}{4\pi\varepsilon_0}q_3\left(\frac{q_1}{r_{13}} + \frac{q_2}{r_{23}}\right) = -\frac{q}{4\pi\varepsilon_0}\left(\frac{3q}{2a} - \frac{2q}{\sqrt{2}a}\right) = \frac{q^2}{4\pi\varepsilon_0}\left(\frac{2\sqrt{2}-3}{2a}\right).$$

Moving in q_4, we have

$$W_4 = \frac{1}{4\pi\varepsilon_0}q_4\left(\frac{q_1}{r_{14}} + \frac{q_2}{r_{24}} + \frac{q_3}{r_{34}}\right) = \frac{q}{4\pi\varepsilon_0}\left(\frac{3q}{a\sqrt{10}} - \frac{2q}{2a} - \frac{q}{a\sqrt{2}}\right)$$
$$= \frac{q^2}{4\pi\varepsilon_0}\left(\frac{3\sqrt{10} - 5\sqrt{2} - 10}{10a}\right).$$

Therefore,

$$W = W_2 + W_3 + W_4 = \frac{q^2}{4\pi\varepsilon_0}\left(-\frac{3\sqrt{2}}{a} + \frac{2\sqrt{2}-3}{2a} + \frac{3\sqrt{10} - 5\sqrt{2} - 10}{10a}\right)$$
$$W = \frac{q^2}{4\pi\varepsilon_0}\left[\frac{3\sqrt{10} - 25(\sqrt{2}+1)}{10a}\right].$$

Problem 2.15. Find the energy stored in a spherical shell of inner radius a and outer radius b with a charge distribution $\rho = kr^2$.

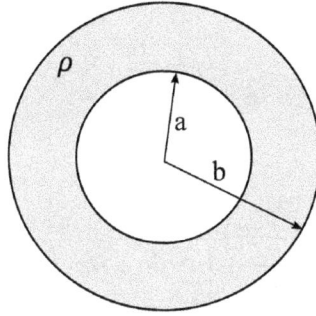

Solution The work is given by

$$W = \frac{\varepsilon_0}{2} \int\limits_{\substack{\text{all} \\ \text{space}}} E^2 \, d\tau$$

so we need to find the field in all three regions.

For $r < a$, we have $q_{enc} = 0$. So $E = 0$.

For $a < r < b$, we have

$$\oint_S \vec{E} \cdot d\vec{a} = \frac{q_{enc}}{\varepsilon_0}$$

with

$$q_{enc} = \int \rho \, d\tau = 4\pi \int\limits_a^r k(r')^2(r')^2 \, dr' = \frac{4\pi k}{5}(r^5 - a^5)$$

and

$$\oint_S \vec{E} \cdot d\vec{a} = E 4\pi r^2.$$

So

$$E = \frac{k(r^5 - a^5)}{5\varepsilon_0 r^2}.$$

For $r > b$, we have

$$q_{enc} = \int \rho \, d\tau = 4\pi \int\limits_a^b k(r')^2(r')^2 \, dr' = \frac{4\pi k}{5}(b^5 - a^5).$$

So

$$E = \frac{k\left(b^5 - a^5\right)}{5\varepsilon_0 r^2}.$$

Now the work is given by

$$W = \frac{\varepsilon_0}{2}\left\{4\pi \int_a^b \left[\frac{k\left(r^5 - a^5\right)}{5\varepsilon_0 r^2}\right]^2 r^2 \, dr + 4\pi \int_b^\infty \left[\frac{k\left(b^5 - a^5\right)}{5\varepsilon_0 r^2}\right]^2 r^2 \, dr\right\}$$

$$W = \frac{k^2\pi}{45\varepsilon_0}\left(5a^9 - 9a^5 b^4 + 4b^9\right).$$

Problem 2.16. Given a charge density $\rho = ke^{-r}$, with k a constant, find the radius of a sphere that maximizes the energy per unit volume.

Solution
The energy per unit volume is given by

$$\frac{W}{\text{volume}} = \frac{\varepsilon_0}{2}E^2,$$

where the field is given by

$$\oint_S \vec{E} \cdot d\vec{a} = \frac{q_{\text{enc}}}{\varepsilon_0}$$

with

$$q_{\text{enc}} = \int \rho \, d\tau = 4\pi \int_0^r ke^{-r'}\left(r'\right)^2 \, dr' = 4\pi ke^{-r}\left(2e^r - r^2 - 2r - 2\right).$$

Also,

$$\oint_S \vec{E} \cdot d\vec{a} = E4\pi r^2.$$

So

$$E = \frac{ke^{-r}\left(2e^r - r^2 - 2r - 2\right)}{\varepsilon_0 r^2}.$$

The energy per unit volume contained in a sphere of radius r is given by

$$\frac{W}{\text{volume}} = \frac{\varepsilon_0}{2}E^2 = \frac{k^2 e^{-2r}\left(2e^r - r^2 - 2r - 2\right)^2}{2\varepsilon_0 r^4}.$$

To maximize this, we have

$$\frac{d}{dr}\left(\frac{W}{volume}\right) = 0$$

so

$$\frac{-k^2 e^{-2r}\left(2e^r - r^2 - 2r - 2\right)\left(4e^r - r^3 - 2r^2 - 4r - 4\right)}{\varepsilon_0 r^5} = 0.$$

Since $r \neq 0$, $k \neq 0$, and $e^{-2r} \neq 0$, we have

$$2e^r - r^2 - 2r - 2 = 0 \rightarrow r = 0$$

and

$$4e^r - r^3 - 2r^2 - 4r - 4 = 0 \rightarrow r = 0, \quad r = 1.45123.$$

But $r \neq 0$, so a sphere of radius $r = 1.45123$ has the maximum energy per unit volume.

Problem 2.17. A metal sphere of radius R and charge q is surrounded by two concentric metal shells.
a) Obtain the surface charge density σ at R, a, b, c, and d.
b) Calculate the potential at the center of the sphere by taking infinity as the reference point.

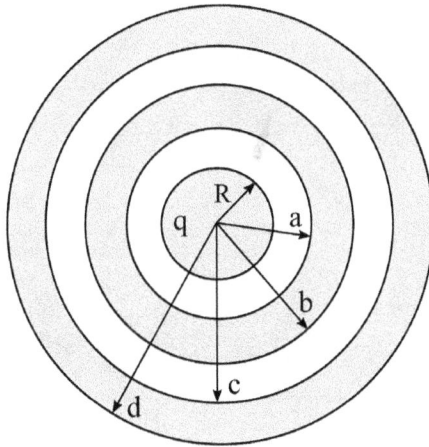

Solutions
a) Obtain the surface charge density σ at R, a, b, c, and d.
   For $r = R$, the sphere is metallic, therefore all charge q is distributed on the surface of the sphere. This gives a surface charge density of

$$\sigma = \frac{q}{4\pi R^2}$$

By influence and due to the sphere with charge q, the inner shell is redistributing the electric charge such that the surface with radius a has charge $-q$ and the surface with radius b has charge $+q$. Similarly for the outer shell.

Therefore, the surface charge densities are

$$\sigma_a = -\frac{q}{4\pi a^2} \qquad \sigma_b = \frac{q}{4\pi b^2} \qquad \sigma_c = -\frac{q}{4\pi c^2} \qquad \sigma_d = \frac{q}{4\pi d^2}.$$

b) Calculate the potential at the center of the sphere by taking infinity as the reference point

Taking our reference point at infinity, the electric potential at the center is given by

$$V = -\int_{\infty}^{0} \vec{E} \cdot d\vec{\ell}$$

$$= \int_{\infty}^{d} \frac{q}{4\pi\varepsilon_0 r^2} dr - \int_{d}^{c} 0 \, dr - \int_{c}^{b} \frac{q}{4\pi\varepsilon_0 r^2} dr - \int_{b}^{a} 0 \, dr - \int_{a}^{R} \frac{q}{4\pi\varepsilon_0 r^2} dr - \int_{R}^{0} 0 \, dr$$

$$V = \frac{q}{4\pi\varepsilon_0}\left(\frac{1}{d} + \frac{1}{b} - \frac{1}{c} + \frac{1}{R} - \frac{1}{a}\right).$$

Problem 2.18. Calculate the capacitance of the spherical shell capacitor of radii a (inner) and b (outer) shown below.

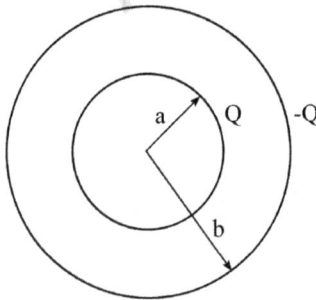

Solution The electric field is due to the inner charge, so

$$E = \frac{Q}{4\pi\varepsilon_0 r^2}.$$

The electric potential is then

$$V = -\int_a^b \vec{E} \cdot d\vec{\ell} = \int_b^a \frac{Q}{4\pi\varepsilon_0 r^2} dr = \frac{Q}{4\pi\varepsilon_0}\left(\frac{1}{a} - \frac{1}{b}\right) = \frac{Q}{4\pi\varepsilon_0}\frac{(b-a)}{ab}.$$

We can find capacitance using

$$V = \frac{Q}{C} \rightarrow C = \frac{Q}{V}.$$

So,

$$C = \frac{Q}{\dfrac{Q}{4\pi\varepsilon_0}\dfrac{(b-a)}{ab}} = \frac{4\pi\varepsilon_0 ab}{b-a}.$$

Problem 2.19. Calculate the capacitance of a cylindrical capacitor of length L with two metal cylinders of radii a (inner) and b (outer) shown below. Ignore the edge effects, obtain the capacitance per unit length.

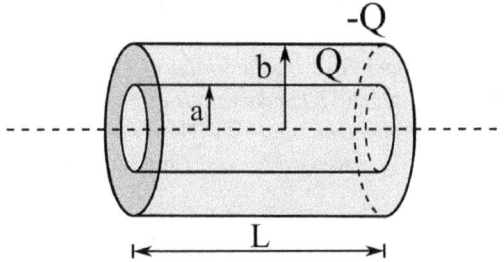

Solution Let us consider that the charge on the inner cylinder is Q (at the radius a). The electric field is obtained from Gauss's law

$$\oint_S \vec{E} \cdot d\vec{a} = \frac{q_{enc}}{\varepsilon_0},$$

where

$$\oint_S \vec{E} \cdot d\vec{a} = \oint_S E\, da = E \oint_S da = E 2\pi s L$$

and the enclosed charge is

$$q_{enc} = Q.$$

Therefore,

$$\vec{E} = \frac{Q}{2\pi\varepsilon_0 sL}\hat{s}.$$

The potential difference between the two cylinders is then

$$V(b) - V(a) = -\int_a^b \vec{E} \cdot d\vec{\ell} = -\int_a^b \frac{Q}{2\pi\varepsilon_0 sL}ds = \frac{Q}{2\pi\varepsilon_0 L}\ln\frac{b}{a}.$$

Therefore, the capacitance is given by

$$C = \frac{Q}{V} = \frac{2\pi\varepsilon_0 L}{\ln\frac{b}{a}}.$$

Bibliography

Byron F W and Fuller R W 1992 *Mathematics of Classical and Quantum Physics* (New York: Dover)

Griffiths D J 1999 *Introduction to Electrodynamics* 3rd edn (Englewood Cliffs, NJ: Prentice Hall)

Griffiths D J 2013 *Introduction to Electrodynamics* 4th edn (New York: Pearson)

Halliday D, Resnick R and Walker J 2010 *Fundamentals of Physics* 9th edn (New York: Wiley)

Halliday D, Resnick R and Walker J 2013 *Fundamentals of Physics* 10th edn (New York: Wiley)

Jackson J D 1998 *Classical Electrodynamics* 3rd edn (New York: Wiley)

Rogawski J 2011 *Calculus: Early Transcendentals* 2nd edn (San Fransisco, CA: Freeman)

Electromagnetism
Problems and solutions
Carolina C Ilie and Zachariah S Schrecengost

Chapter 3

Electric potential

Chapter 3 contains different methods for obtaining the electric potential. We will focus on calculating the potential as finding the field is a straightforward calculation once the potential has been determined. Laplace's equation is solved using different methods, depending on the type of charge distribution and on the symmetry of the problem. The method of images, separation of variables, and multipole (in particular dipole) expansions are discussed using appropriate examples.

3.1 Theory

3.1.1 Laplace's equation

Cartesian

$$\nabla^2 T = \frac{\partial^2 T}{\partial x^2} + \frac{\partial^2 T}{\partial y^2} + \frac{\partial^2 T}{\partial z^2} = 0$$

Cylindrical

$$\nabla^2 T = \frac{1}{s}\frac{\partial}{\partial s}\left(s\frac{\partial T}{\partial s}\right) + \frac{1}{s^2}\frac{\partial^2 T}{\partial \phi^2} + \frac{\partial^2 T}{\partial z^2} = 0$$

Spherical

$$\nabla^2 T = \frac{1}{r^2}\frac{\partial}{\partial r}\left(r^2\frac{\partial T}{\partial r}\right) + \frac{1}{r^2 \sin\theta}\frac{\partial}{\partial \theta}\left(\sin\theta \frac{\partial T}{\partial \theta}\right) + \frac{1}{r^2 \sin^2\theta}\frac{\partial^2 T}{\partial \phi^2} = 0$$

3.1.2 Solving Laplace's equation

As an introduction to solving problems using Laplace's equation, we will outline the solutions in Cartesian and spherical coordinates. Laplace's equation can be solved by the method separation of variables when we know the boundary conditions. The

doi:10.1088/978-1-6817-4429-2ch3

general solutions will be outlined below, but seeing how they are derived is important. We will leave them in a general form and problems in this chapter will provide examples of using the boundary conditions to solve for the constants.

Two-dimensional Cartesian coordinates
Let us look at a general two-dimensional case where Laplace's equation is given by

$$\nabla^2 V = \frac{\partial^2 V}{\partial x^2} + \frac{\partial^2 V}{\partial y^2} = 0.$$

We look for a solution of the type $V(x, y) = X(x)Y(y)$ and we replace the desired solution in Laplace's equation which becomes

$$Y(y)\frac{\partial^2 X(x)}{\partial x^2} + X(x)\frac{\partial^2 Y(y)}{\partial y^2} = 0$$

and in a simpler form

$$Y\frac{\partial^2 X}{\partial x^2} + X\frac{\partial^2 Y}{\partial y^2} = 0.$$

We want to separate the variables, which can easily be done by dividing the equation by $X(x)Y(y) = V(x, y)$

$$\frac{1}{X}\frac{\partial^2 X}{\partial x^2} + \frac{1}{Y}\frac{\partial^2 Y}{\partial y^2} = 0.$$

Note that the first term depends only on x and the second term depends only on y. This means that each of the two terms must be constant, and the two constants must be equal in magnitude but opposite in sign. So

$$\frac{1}{X}\frac{\partial^2 X}{\partial x^2} = F \quad \frac{1}{Y}\frac{\partial^2 Y}{\partial y^2} = -F.$$

We choose F positive, and we can rewrite the equations as

$$\frac{1}{X}\frac{d^2 X}{dx^2} = k^2 \quad \frac{1}{Y}\frac{d^2 Y}{dy^2} = -k^2.$$

Note that the initial partial differential equation was replaced by two ordinary differential equations. Rearranging yields

$$\frac{d^2 X}{dx^2} = k^2 X \quad \frac{d^2 Y}{dy^2} = -k^2 Y.$$

The two equations have the following solutions

$$X(x) = Ae^{kx} + Be^{-kx}$$

and

$$Y(y) = C \sin(ky) + D \cos(ky).$$

Going back to the electric potential, V becomes

$$V(x, y) = X(x)Y(y) = \left(Ae^{kx} + Be^{-kx} \right)\left[C \sin(ky) + D \cos(ky) \right].$$

The next step is to apply the boundary conditions in order to obtain the constants A, B, C, and D and to (usually) impose some constraints on k.

Two-dimensional spherical coordinates
Here we will assume azimuthal symmetry (no dependence on ϕ) where Laplace's equation is given by

$$\nabla^2 V = \frac{1}{r^2}\frac{\partial}{\partial r}\left(r^2 \frac{\partial V}{\partial r} \right) + \frac{1}{r^2 \sin \theta}\frac{\partial}{\partial \theta}\left(\sin \theta \frac{\partial V}{\partial \theta} \right) = 0.$$

We look for a solution which has a radial component and an angular component

$$V(r, \theta) = R(r)\Theta(\theta).$$

Note that here R is the function of r, and not merely the radius of the sphere. We plug our solution in the previous equation and we obtain

$$\Theta(\theta)\frac{\partial}{\partial r}\left(r^2 \frac{\partial R(r)}{\partial r} \right) + \frac{R(r)}{\sin \theta}\frac{\partial}{\partial \theta}\left(\sin \theta \frac{\partial \Theta(\theta)}{\partial \theta} \right) = 0.$$

We want to use the method of separation of variables so we will divide the previous equation by $V(r, \theta) = R(r)\Theta(\theta)$,

$$\frac{1}{R(r)}\frac{d}{dr}\left(r^2 \frac{dR(r)}{dr} \right) + \frac{1}{\Theta(\theta)\sin \theta}\frac{d}{d\theta}\left(\sin \theta \frac{d\Theta(\theta)}{d\theta} \right) = 0.$$

Note that each term is only a function of a single variable so we were able to replace the partial derivates with ordinary derivates. Now we have one term in $R(r)$ and another term in $\Theta(\theta)$, so we have separated the variables. Therefore, each term must be constant. For well know reasons (more apparent in quantum mechanics), we choose the constant as following

$$\frac{1}{R(r)}\frac{d}{dr}\left(r^2 \frac{dR(r)}{dr} \right) = l(l + 1)$$

$$\frac{1}{\Theta(\theta)\sin \theta}\frac{d}{d\theta}\left(\sin \theta \frac{d\Theta(\theta)}{d\theta} \right) = -l(l + 1).$$

Now let us analyze each of the equations and find the solution.

The radial equation

$$\frac{d}{dr}\left(r^2\frac{dR(r)}{dr}\right) = l(l+1)R(r)$$

has the general solution, with A and B constants

$$R(r) = Ar^l + \frac{B}{r^{l+1}}.$$

The angular equation

$$\frac{d}{d\theta}\left(\sin\theta\,\frac{d\Theta(\theta)}{d\theta}\right) = -l(l+1)\Theta(\theta)\sin\theta$$

is not at all trivial. The solutions constitute Legendre polynomials with the variable $\cos\theta$. Legendre polynomials are a special class of polynomials. So the solution to the angular equation is

$$\Theta(\theta) = P_l(\cos\theta),$$

where the general form is given by Rodrigues formula

$$P_l(x) = \frac{1}{2^l l!}\left(\frac{d}{dx}\right)^l\left(x^2 - 1\right)^2.$$

Therefore, the separable solution of the Laplace equation, considering azimuthal symmetry, is

$$V(r, \theta) = R(r)\Theta(\theta) = \left(Ar^l + \frac{B}{r^{l+1}}\right)P_l(\cos\theta)$$

and the general solution is the linear combination of the separable solutions

$$V(r, \theta) = \sum_{l=0}^{\infty}\left(A_l r^l + \frac{B_l}{r^{l+1}}\right)P_l(\cos\theta).$$

3.1.3 General solutions

Cartesian

$$\frac{\partial^2 V}{\partial x^2} + \frac{\partial^2 V}{\partial y^2} = 0 \rightarrow V(x, y) = (Ae^{kx} + Be^{-kx})\left[C\sin(ky) + D\cos(ky)\right]$$

Spherical

$$\frac{1}{r^2}\frac{\partial}{\partial r}\left(r^2\frac{\partial V}{\partial r}\right) + \frac{1}{r^2\sin\theta}\frac{\partial}{\partial\theta}\left(\sin\theta\,\frac{\partial V}{\partial\theta}\right) = 0$$

$$V(r, \theta) = \sum_{l=0}^{\infty} \left(A_l r^l + \frac{B_l}{r^{l+1}} \right) P_l(\cos \theta),$$

where P_l are Legendre polynomials given by the Rodrigues formula

$$P_l(x) = \frac{1}{2^l l!} \left(\frac{d}{dx} \right)^l (x^2 - 1)^l.$$

Note

$$P_0(x) = 1$$

$$P_1(x) = x$$

$$P_2(x) = \frac{3x^2 - 1}{2}$$

$$P_3(x) = \frac{5x^3 - 3x}{2}$$

$$P_4(x) = \frac{35x^4 - 30x^2 + 3}{8}.$$

Cylindrical

$$\frac{1}{s} \frac{\partial}{\partial s} \left(s \frac{\partial V}{\partial s} \right) + \frac{1}{s^2} \frac{\partial^2 V}{\partial \phi^2} + \frac{\partial^2 V}{\partial z^2} = 0$$

$$V(s, \phi) = a_0 + b_0 \ln(s) + \sum_{k=1}^{\infty} \left\{ s^k \left[a_k \cos(k\phi) + b_k \sin(k\phi) \right] \right.$$
$$\left. + s^{-k} \left[c_k \cos(k\phi) + d_k \sin(k\phi) \right] \right\}.$$

3.1.4 Method of images

The method of images is a very useful technique for calculating the electric field and the electric potential for problems with symmetry. By using the uniqueness theorem we know that the electric field is uniquely determined at any point in space, thus we can replace an apparently difficult problem with another problem in which we use the initial charge(s) and also the 'images' of the charges. The boundary conditions need to be fulfilled. Typically, this involves the condition for a zero electric potential for a grounded conductor and the condition for zero potential very far away from the system of charges. The potential can be determined in the permitted region, which is in general the region of the real charge. The region of the image charges is the 'forbidden' region; the potential cannot be calculated there. The best way to learn this method is by solving problems and checking the examples.

3.1.5 Potential due to a dipole

$$V(\vec{r}) \cong \frac{1}{4\pi\varepsilon_0} \frac{qd\cos\theta}{r^2} = \frac{1}{4\pi\varepsilon_0} \frac{\vec{p} \cdot \hat{r}}{r^2}$$

3.1.6 Multiple expansion

$$V(\vec{r}) = \frac{1}{4\pi\varepsilon_0} \sum_{n=0}^{\infty} \frac{1}{r^{n+1}} \int (r')^n P_n(\cos\theta') \, \rho(\vec{r}') \mathrm{d}\tau'$$

$$V(\vec{r}) = \frac{1}{4\pi\varepsilon_0} \left[\frac{1}{r} \int \rho(\vec{r}') \mathrm{d}\tau' + \frac{1}{r^2} \int r' \cos\theta' \, \rho(\vec{r}') \mathrm{d}\tau' \right.$$

$$\left. + \frac{1}{r^3} \int (r')^2 \left(\frac{3}{2} \cos^2\theta' - \frac{1}{2} \right) \rho(\vec{r}') \mathrm{d}\tau' + \cdots \right].$$

3.1.7 Monopole moment

$$Q = \sum_{i=0}^{n} q_i$$

3.2 Problems and solutions

Problem 3.1. Solve the Laplace equation in spherical and cylindrical coordinates for the cases where V is only dependent on one coordinate at a time.

Solution In spherical coordinates

$$\nabla^2 V = \frac{1}{r^2} \frac{\partial}{\partial r} \left(r^2 \frac{\partial V}{\partial r} \right) + \frac{1}{r^2 \sin\theta} \frac{\partial}{\partial \theta} \left(\sin\theta \, \frac{\partial V}{\partial \theta} \right) + \frac{1}{r^2 \sin^2\theta} \frac{\partial^2 V}{\partial \phi^2} = 0.$$

If V only depends on r

$$\frac{1}{r^2} \frac{\mathrm{d}}{\mathrm{d}r} \left(r^2 \frac{\mathrm{d}V}{\mathrm{d}r} \right) = 0,$$

which means

$$r^2 \frac{\mathrm{d}V}{\mathrm{d}r} = C.$$

So

$$V = C \int r^{-2} \, \mathrm{d}r \rightarrow V(r) = C\left(-r^{-1} + A \right) = k - \frac{C}{r}.$$

If V only depends on θ

$$\frac{1}{r^2 \sin \theta} \frac{\mathrm{d}}{\mathrm{d}\theta}\left(\sin \theta \, \frac{\mathrm{d}V}{\mathrm{d}\theta}\right) = 0,$$

which mean

$$\sin \theta \, \frac{\mathrm{d}V}{\mathrm{d}\theta} = C.$$

So

$$V = C \int \frac{1}{\sin \theta} \mathrm{d}\theta = C \int \csc \theta \, \mathrm{d}\theta = C\left(\ln |\csc \theta - \cot \theta| + A\right)$$

$$V(\theta) = k + C \ln |\csc \theta - \cot \theta|.$$

If V only depends on ϕ

$$\frac{1}{r^2 \sin^2 \theta} \frac{\mathrm{d}^2 V}{\mathrm{d}\phi^2} = 0,$$

which means

$$\frac{\mathrm{d}V}{\mathrm{d}\phi} = C.$$

So

$$V(\phi) = k + C\phi.$$

In cylindrical coordinates

$$\frac{1}{s} \frac{\partial}{\partial s}\left(s \frac{\partial V}{\partial s}\right) + \frac{1}{s^2} \frac{\partial^2 V}{\partial \phi^2} + \frac{\partial^2 V}{\partial z^2} = 0.$$

If V only depends on s

$$\frac{1}{s} \frac{\mathrm{d}}{\mathrm{d}s}\left(s \frac{\mathrm{d}V}{\mathrm{d}s}\right) = 0,$$

which means

$$s \frac{\mathrm{d}V}{\mathrm{d}s} = C.$$

So

$$V = C \int s^{-1} \, \mathrm{d}s = C\left(\ln |s| + A\right)$$

$$V(s) = k + C \ln |s|.$$

If V only depends on ϕ,

$$\frac{1}{s^2}\frac{d^2V}{d\phi^2} = 0,$$

which means

$$\frac{dV}{d\phi} = C,$$

which is the same as ϕ dependence in spherical coordinates. So

$$V(\phi) = k + C\phi.$$

If V only depends on z

$$\frac{d^2V}{dz^2} = 0,$$

which means

$$\frac{dV}{dz} = C,$$

which is the same form as ϕ dependence. So

$$V(z) = k + Cz.$$

Problem 3.2. In two-dimensional Cartesian coordinates, the general solution to the Laplace equation is

$$V(x, y) = \left(Ae^{kx} + Be^{-kx}\right)\left[C \sin(ky) + D \cos(ky)\right].$$

Verify that this does in fact satisfy the Laplace equation.

Solution Here, the Laplace equation is

$$\frac{\partial^2 V}{\partial x^2} + \frac{\partial^2 V}{\partial y^2} = 0.$$

So we need to compute $\frac{\partial^2 V}{\partial x^2}$ and $\frac{\partial^2 V}{\partial y^2}$. We have

$$\frac{\partial V}{\partial x} = \left(Ake^{kx} - Bke^{-kx}\right)\left[C \sin(ky) + D \cos(ky)\right]$$

and

$$\frac{\partial^2 V}{\partial x^2} = \left(Ak^2e^{kx} + Bk^2e^{-kx}\right)\left[C \sin(ky) + D \cos(ky)\right]$$

$$= k^2\left(Ae^{kx} + Be^{-kx}\right)\left[C \sin(ky) + D \cos(ky)\right].$$

Also,

$$\frac{\partial V}{\partial y} = \left(Ae^{kx} + Be^{-kx}\right)\left[Ck\cos(ky) - Dk\sin(ky)\right]$$

and

$$\frac{\partial^2 V}{\partial y^2} = \left(Ae^{kx} + Be^{-kx}\right)\left[-Ck^2\sin(ky) - Dk^2\cos(ky)\right]$$

$$= -k^2\left(Ae^{kx} + Be^{-kx}\right)\left[C\sin(ky) + D\cos(ky)\right].$$

Putting this all together, we have

$$\frac{\partial^2 V}{\partial x^2} + \frac{\partial^2 V}{\partial y^2} = k^2\left(Ae^{kx} + Be^{-kx}\right)\left[C\sin(ky) + D\cos(ky)\right]$$

$$- k^2\left(Ae^{kx} + Be^{-kx}\right)\left[C\sin(ky) + D\cos(ky)\right]$$

$$\frac{\partial^2 V}{\partial x^2} + \frac{\partial^2 V}{\partial y^2} = 0$$

as expected.

Problem 3.3. A charge q is placed at a distance d from an infinite grounded conducting plane. Using the method of images, find the electric potential. Which is the 'forbidden' region, for which we cannot calculate the potential?

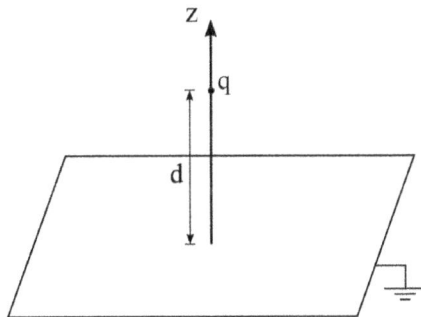

Solution We replace the previous problem with a completely different problem: charge q at $(0, 0, d)$ and its image, charge $-q$ situated at $(0, 0, -d)$. The grounded conducting plane disappeared. This new problem is shown below

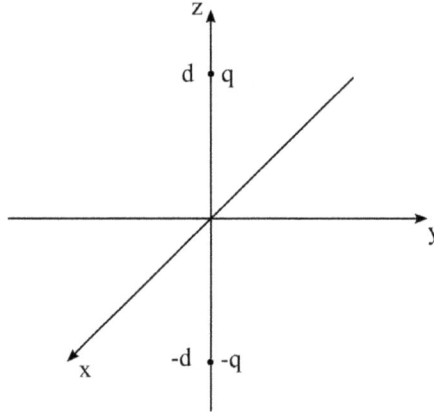

The potential needs to satisfy the following boundary conditions:
a) $V = 0$ for $z = 0$ (grounded plane in the initial problem).
b) $V \to 0$ for a point far from charge q

$$x^2 + y^2 + z^2 \gg d^2.$$

The electric potential due to both point charges is:

$$V(x, y, z) = \frac{q}{4\pi\varepsilon_0\sqrt{x^2 + y^2 + (z - d)^2}} + \frac{-q}{4\pi\varepsilon_0\sqrt{x^2 + y^2 + (z + d)^2}}.$$

Let us check the boundary conditions:
a) For $z = 0$ it is easy to see that $V = 0$.

b) For $x^2 + y^2 + z^2 \gg d^2$, $\sqrt{x^2 + y^2 + (z - d)^2} \cong \sqrt{x^2 + y^2 + (z + d)^2}$, and $V \to 0$.

It is important to note that the only region for which we are able to obtain the electric potential is the region in space above the grounded conducting plane, i.e. the semi-space where charge q is located. For $z < 0$, we are not able to obtain the electric potential.

Problem 3.4. A charge q is placed in an opened grounded conducting parallelepiped at (a, b, c), where a, b, c are positive. Using the method of images, obtain the electric potential in the region of the charge q (for which $x > 0$, $y > 0$, $z > 0$).

3-10

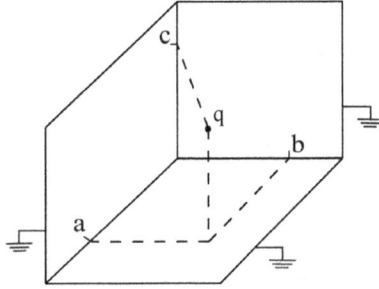

Solution We have the one real charge q at (a, b, c) and seven image charges as following: $-q$ at $(-a, -b, -c)$, $(-a, b, c)$, $(a, -b, c)$, and $(a, b, -c)$; q at $(a, -b, -c)$, $(-a, b, -c)$, $(-a, -b, c)$. The electric potential is given by

$$V(x, y, z)$$

$$= \frac{1}{4\pi\varepsilon_0}\left(\frac{q}{\sqrt{(x-a)^2 + (y-b)^2 + (z-c)^2}} + \frac{q}{\sqrt{(x-a)^2 + (y+b)^2 + (z+c)^2}}\right.$$

$$+ \frac{q}{\sqrt{(x+a)^2 + (y-b)^2 + (z+c)^2}} + \frac{q}{\sqrt{(x+a)^2 + (y+b)^2 + (z-c)^2}}$$

$$+ \frac{-q}{\sqrt{(x+a)^2 + (y+b)^2 + (z+c)^2}} + \frac{-q}{\sqrt{(x-a)^2 + (y+b)^2 + (z-c)^2}}$$

$$\left. + \frac{-q}{\sqrt{(x-a)^2 + (y-b)^2 + (z+c)^2}} + \frac{-q}{\sqrt{(x+a)^2 + (y-b)^2 + (z-c)^2}} \right).$$

If we check the limit $x = 0$, $y = 0$, $z = 0$ successively, we obtain a zero potential for all the three sides of the parallelepiped, where the grounded conductors were in the equivalent problem. Also, for the points far away from the point (a, b, c) in the 'eighth' part of space for which $x \gg 0$, $y \gg 0$, $z \gg 0$, the potential becomes zero as well. Again, note that this is the electric potential only for this part of space, accessible for investigation using the method of images.

Problem 3.5. Let us imagine that we have n charges placed as following: $q_1 = -q$ at $(0, 0, d)$, $q_2 = 2q$ at $(0, 0, 2d)$, ..., $q_n = (-1)^n nq$ at $(0, 0, nd)$ above a grounded, conducting xy-plane (shown below). Obtain the electric potential using the method of images.

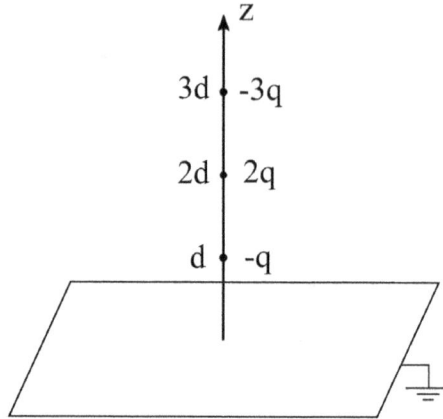

Solution The image charges will be equal in magnitude, of different sign, and situated symmetrically with the xy-plane. In the new problem we eliminate the grounded, conducting plane, but we use the xy-plane for geometrical purposes.

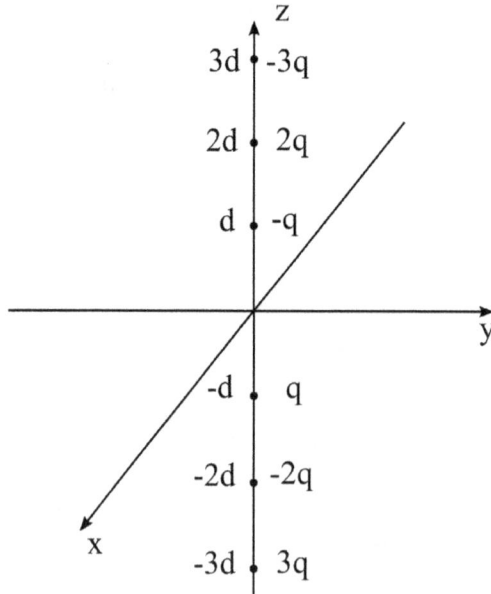

For each charge $q_n = (-1)^n nq$ we have the image charge $q_n' = (-1)^{n+1} nq$ located at $(0, 0, -nd)$. The electric potential is, therefore,

$$V(x, y, z) = \frac{q}{4\pi\varepsilon_0} \left(\frac{-1}{\sqrt{x^2 + y^2 + (z-d)^2}} + \frac{1}{\sqrt{x^2 + y^2 + (z+d)^2}} \right.$$

$$+ \frac{(-1)^2 2}{\sqrt{x^2 + y^2 + (z-2d)^2}} + \frac{(-1)^{2+1} 2}{\sqrt{x^2 + y^2 + (z+2d)^2}} + \cdots$$

$$+ \frac{(-1)^k k}{\sqrt{x^2 + y^2 + (z-kd)^2}} + \frac{(-1)^{k+1} k}{\sqrt{x^2 + y^2 + (z+kd)^2}}$$

$$+ \cdots + \frac{(-1)^n n}{\sqrt{x^2 + y^2 + (z-nd)^2}} + \left. \frac{(-1)^{n+1} n}{\sqrt{x^2 + y^2 + (z+nd)^2}} \right).$$

It is easy to see that $V = 0$ for $z = 0$ and also $V = 0$ for a point very far from charge $x^2 + y^2 + z^2 \gg (nd)^2$.

Problem 3.6. A conducting sphere of radius R, centered at the origin, is grounded. Find the potential outside the sphere, if a point charge $+q$ is placed at a distance d from the sphere, $d > R$. Use the method of images.

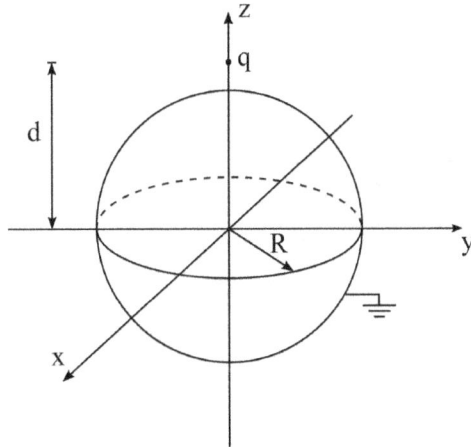

Solution We replace our problem with the grounded, conducting sphere of radius R and the charge $+q$ at distance $d > R$ with a different problem. The sphere, the charge $+q$ and the image charge q', situated at $(0, 0, a)$, with $a < R$, is given by

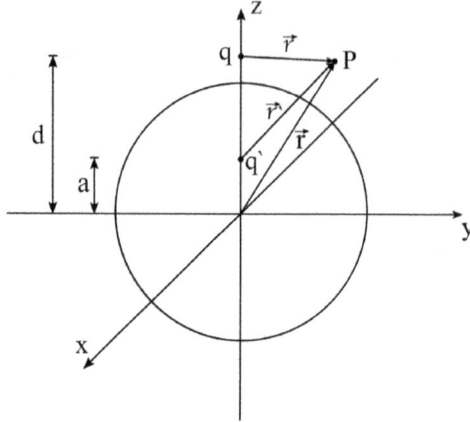

We need $V = 0$ everywhere on the sphere of radius R. Note that we can only find the electric potential outside the sphere. Consider the point P depicted above; here the electric potential at P is given by

$$V(\vec{r}) = \frac{1}{4\pi\varepsilon_0}\left(\frac{q}{r_1} + \frac{q'}{r_2}\right) = \frac{1}{4\pi\varepsilon_0}\left(\frac{q}{|\vec{r} - \vec{d}|} + \frac{q'}{|\vec{r} - \vec{a}|}\right).$$

Using the law of cosines, we have

$$r_1^2 = d^2 + r^2 - 2dr\cos\theta$$

and

$$r_2^2 = a^2 + r^2 - 2ar\cos\theta.$$

We can rewrite V as

$$V(\vec{r}) = \frac{1}{4\pi\varepsilon_0}\left(\frac{q}{\sqrt{d^2 + r^2 - 2dr\cos\theta}} + \frac{q'}{\sqrt{a^2 + r^2 - 2ar\cos\theta}}\right).$$

The potential should be zero for $r = R$

$$V(R) = \frac{1}{4\pi\varepsilon_0}\left(\frac{q}{\sqrt{d^2 + R^2 - 2dR\cos\theta}} + \frac{q'}{\sqrt{a^2 + R^2 - 2aR\cos\theta}}\right) = 0.$$

So

$$\frac{q}{\sqrt{d^2 + R^2 - 2Rd\cos\theta}} = \frac{-q'}{\sqrt{a^2 + R^2 - 2aR\cos\theta}}.$$

We need to obtain both q' and a, so we need two equations. We choose two convenient values for θ, $\theta = 0$ and $\theta = \pi$. For $\theta = 0$, $\cos\theta = 1$, so

$$\frac{q}{\sqrt{d^2 + R^2 - 2Rd}} = \frac{-q'}{\sqrt{a^2 + R^2 - 2aR}} \rightarrow \frac{q}{\sqrt{(d - R)^2}} = \frac{-q'}{\sqrt{(R - a)^2}}.$$

Choosing the positive square root,

$$\frac{q}{d - R} = \frac{-q'}{R - a}$$

and solving for the image charge, we obtain

$$q' = -q\frac{R - a}{d - R}.$$

For $\theta = \pi$, $\cos \theta = -1$, so

$$\frac{q}{d + R} = \frac{-q'}{R + a}.$$

By substituting q', we have

$$\frac{q}{d + R} = \frac{(R - a)q}{(d - R)(a + R)}.$$

Solving for a, we obtain

$$a = \frac{R^2}{d}.$$

Using this, we can find our image charge

$$q' = -\frac{\left(R - \dfrac{R^2}{d}\right)q}{d - R} = -\frac{R(d - R)q}{(d - R)d}$$

$$q' = -\frac{R}{d}q.$$

Now we have the electric potential, since we obtained the image charge and its position.

Problem 3.7. Given two infinitely long grounded plates at $y = 0$ and $y = a$ connected by the metal strip at $x = -b$ with constant potential $-V_0$ and $x = b$ with constant potential V_0. Find the potential inside the pipe.

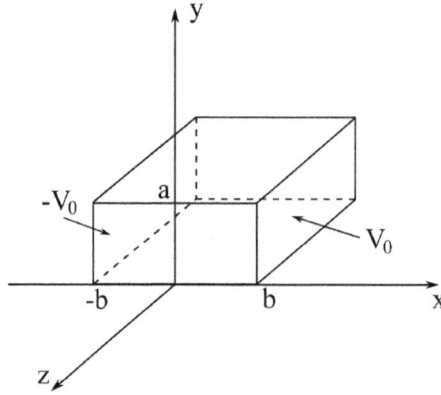

Solution This is independent of z so we have

$$\frac{\partial^2 V}{\partial x^2} + \frac{\partial^2 V}{\partial y^2} = 0$$

with boundary conditions
  (i) $V(y = 0) = 0$
 (ii) $V(y = a) = 0$
(iii) $V(x = b) = V_0$
(iv) $V(x = -b) = -V_0$.

Our general solution is given by

$$V(x, y) = \left(Ae^{kx} + Be^{-kx}\right)\left[C \sin(ky) + D \cos(ky)\right].$$

From boundary condition (i)

$$V(x, 0) = \left(Ae^{kx} + Be^{-kx}\right)(D) = 0 \Rightarrow D = 0.$$

So our solution becomes

$$V(x, y) = \left(Ae^{kx} + Be^{-kx}\right)\left[C \sin(ky)\right].$$

From boundary condition (ii)

$$V(x, y) = \left(Ae^{kx} + Be^{-kx}\right)\left[C \sin(ka)\right] = 0$$

we have

$$k = \frac{n\pi}{a},$$

where n is an integer. By symmetry

$$V(x, y) = -V(-x, y)$$

$$\left(Ae^{kx} + Be^{-kx}\right)[C \sin(ka)] = -\left(Ae^{-kx} + Be^{kx}\right)[C \sin(ka)]$$

$$Ae^{kx} + Be^{-kx} = -Ae^{-kx} + Be^{kx}$$

$$A\left(e^{kx} + e^{-kx}\right) = -B\left(e^{kx} + e^{-kx}\right).$$

So

$$A = -B.$$

Absorbing our constants, the solution becomes

$$V(x, y) = C\left(e^{kx} + e^{-kx}\right)\sin(ky) = C \sinh(kx)\sin(ky)$$

and in general

$$V(x, y) = \sum_{n=1}^{\infty} C_n \sinh\left(\frac{n\pi x}{a}\right)\sin\left(\frac{n\pi y}{a}\right).$$

Now to find C_n we take

$$V(b, y) = \sum_{n=1}^{\infty} C_n \sinh\left(\frac{n\pi b}{a}\right)\sin\left(\frac{n\pi y}{a}\right) = V_0$$

so

$$\sum_{n=1}^{\infty} C_n \sinh\left(\frac{n\pi b}{a}\right) \int_0^a \sin\left(\frac{n\pi y}{a}\right)\sin\left(\frac{n'\pi y}{a}\right)dy = \int_0^a V_0 \sin\left(\frac{n'\pi y}{a}\right)dy.$$

Note that when $n \neq n'$,

$$\int_0^a \sin\left(\frac{n\pi y}{a}\right)\sin\left(\frac{n'\pi y}{a}\right)dy = 0$$

and when $n = n'$,

$$\int_0^a \sin\left(\frac{n\pi y}{a}\right)\sin\left(\frac{n'\pi y}{a}\right)dy = \frac{a}{2}.$$

Therefore,

$$C_n \sinh\left(\frac{n\pi b}{a}\right)\frac{a}{2} = \frac{V_0 a}{n\pi}[1 - \cos(n\pi)] = \begin{cases} 0 & n \text{ is even} \\ \dfrac{2V_0 a}{n\pi} & n \text{ is odd} \end{cases}.$$

So

$$C_n = \frac{4V_0}{n\pi} \frac{1}{\sinh\left(\frac{n\pi b}{a}\right)}$$

for $n = 1, 3, 5, \ldots$ and our potential is given by

$$V(x, y) = \frac{4V_0}{\pi} \sum_{n=1,3,5}^{\infty} \frac{1}{n} \frac{\sinh\left(\frac{n\pi x}{a}\right)}{\sinh\left(\frac{n\pi b}{a}\right)} \sin\left(\frac{n\pi y}{a}\right).$$

Problem 3.8. Suppose a thin spherical shell of radius R has potential $V(\theta) = V_0(1 - \frac{3}{2}\sin^2\theta)$ specified at the surface. Find the potential inside and outside the sphere.

Solution Our general solution is given by

$$V(r, \theta) = \sum_{l=0}^{\infty}\left(A_l r^l + \frac{B_l}{r^{l+1}}\right)P_l(\cos\theta).$$

Inside:
Here, we must have $B_l = 0$ so the potential does not blow up at the origin. So our potential becomes

$$V(r, \theta) = \sum_{l=0}^{\infty}A_l r^l P_l(\cos\theta).$$

At the surface, we have

$$V(R, \theta) = \sum_{l=0}^{\infty}A_l R^l P_l(\cos\theta) = V_0\left(1 - \frac{3}{2}\sin^2\theta\right).$$

Note that

$$V_0\left(1 - \frac{3}{2}\sin^2\theta\right) = V_0\left[1 - \frac{3}{2}(1 - \cos^2\theta)\right] = V_0\left(\frac{3\cos^2\theta - 1}{2}\right) = V_0 P_2(\cos\theta).$$

So

$$\sum_{l=0}^{\infty}A_l R^l P_l(\cos\theta) = V_0 P_2(\cos\theta).$$

This means we only have the $l = 2$ term,

$$A_2 R^2 P_2(\cos\theta) = V_0 P_2(\cos\theta) \rightarrow A_2 = \frac{V_0}{R^2}.$$

Therefore,

$$V(r, \theta) = A_2 r^2 P_2(\cos\theta) = V_0\left(\frac{r}{R}\right)^2\left(\frac{3\cos^2\theta - 1}{2}\right).$$

Outside:

Here, we must have $A_l = 0$ so the potential does not blow up as $r \to \infty$. So our potential becomes

$$V(r, \theta) = \sum_{l=0}^{\infty} \frac{B_l}{r^{l+1}} P_l(\cos \theta).$$

At the surface, we have

$$V(R, \theta) = \sum_{l=0}^{\infty} \frac{B_l}{R^{l+1}} P_l(\cos \theta) = V_0 \left(1 - \frac{3}{2} \sin^2 \theta \right).$$

Again, the right-hand side is $V_0 P_2(\cos \theta)$, so

$$\frac{B_2}{R^{2+1}} P_2(\cos \theta) = V_0 P_2(\cos \theta) \rightarrow B_2 = V_0 R^3.$$

Therefore,

$$V(r, \theta) = \frac{B_2}{r^3} P_2(\cos \theta) = V_0 \left(\frac{R}{r} \right)^3 \left(\frac{3 \cos^2 \theta - 1}{2} \right).$$

Problem 3.9. A spherical shell of radius R has surface charge $\sigma_0(\theta) = \sin \theta \sin 3\theta$ smeared on its surface. Find the potential inside and outside the sphere.

Solution Our general solution is given by

$$V(r, \theta) = \sum_{l=0}^{\infty} \left(A_l r^l + \frac{B_l}{r^{l+1}} \right) P_l(\cos \theta).$$

Inside we must have $B_l = 0$, otherwise $V \to \infty$ as $r \to 0$. So our potential becomes

$$V_{\text{in}}(r, \theta) = \sum_{l=0}^{\infty} A_l r^l P_l(\cos \theta).$$

Outside we must have $A_l = 0$, otherwise $V \to \infty$ as $r \to \infty$. So our potential becomes

$$V_{\text{out}}(r, \theta) = \sum_{l=0}^{\infty} \frac{B_l}{r^{l+1}} P_l(\cos \theta).$$

At the surface they must be equal, so

$$V_{\text{in}}(R, \theta) = V_{\text{out}}(R, \theta)$$

$$\sum_{l=0}^{\infty} A_l R^l P_l(\cos \theta) = \sum_{l=0}^{\infty} \frac{B_l}{R^{l+1}} P_l(\cos \theta)$$

$$B_l = A_l R^{2l+1}.$$

We must also have

$$\left(\frac{\partial V_{\text{out}}}{\partial r} - \frac{\partial V_{\text{in}}}{\partial r}\right)\bigg|_{r=R} = -\frac{1}{\epsilon_0}\sigma_0(\theta),$$

where

$$\frac{\partial V_{\text{out}}}{\partial r} = \sum_{l=0}^{\infty} -(l+1)\frac{B_l}{r^{l+2}}P_l(\cos\theta)$$

and

$$\frac{\partial V_{\text{in}}}{\partial r} = \sum_{l=0}^{\infty} lA_l r^{l-1}P_l(\cos\theta).$$

Thus,

$$\left(\frac{\partial V_{\text{out}}}{\partial r} - \frac{\partial V_{\text{in}}}{\partial r}\right)\bigg|_{r=R} = \sum_{l=0}^{\infty} -(l+1)\frac{B_l}{R^{l+2}}P_l(\cos\theta) - lA_l R^{l-1}P_l(\cos\theta).$$

Substitution of B_l yields

$$\left(\frac{\partial V_{\text{out}}}{\partial r} - \frac{\partial V_{\text{in}}}{\partial r}\right)\bigg|_{r=R} = \sum_{l=0}^{\infty} (2l+1)A_l R^{l-1}P_l(\cos\theta) = \frac{1}{\epsilon_0}\sigma_0(\theta).$$

Since Legendre polynomials are orthogonal, when $l \neq l'$ we have

$$\int_0^{\pi} P_l(\cos\theta)P_{l'}(\cos\theta)\sin\theta\,d\theta = \frac{2}{2l+1}.$$

It follows that

$$A_l = \frac{1}{2\epsilon_0 R^{l-1}}\int_0^{\pi} \sigma_0(\theta)P_l(\cos\theta)\sin\theta\,d\theta.$$

Note that $\sigma_0(\theta)$ can be rewritten as

$$\sigma_0(\theta) = \sin\theta \sin 3\theta = \sin\theta(\sin 2\theta \cos\theta + \cos 2\theta \sin\theta)$$

$$= \sin\theta\left[(2\sin\theta\cos\theta)\cos\theta + \sin\theta\,(2\cos^2\theta - 1)\right]$$

$$= 2\sin^2\theta\cos^2\theta + \sin^2\theta\,(2\cos^2\theta - 1)$$

$$= 2(1 - \cos^2\theta)\cos^2\theta + (1 - \cos^2\theta)(2\cos^2\theta - 1)$$

$$= 2\cos^2\theta - 2\cos^4\theta + 2\cos^2\theta - 2\cos^4\theta - 1 + \cos^2\theta$$

$$= -4\cos^4\theta + 5\cos^2\theta - 1.$$

We can find α, β, and γ such that

$$-4\cos^4\theta + 5\cos^2\theta - 1 = \alpha P_4(\cos\theta) + \beta P_2(\cos\theta) + \gamma P_0(\cos\theta).$$

So,

$$-4\cos^4\theta + 5\cos^2\theta - 1 = \alpha\frac{35\cos^4\theta - 30\cos^2\theta + 3}{8} + \beta\frac{3\cos^2\theta - 1}{2} + \gamma.$$

It follows that $\alpha = -\frac{32}{35}$, $\beta = \frac{22}{21}$, and $\gamma = -\frac{2}{15}$ and we can now solve for A_l.

$$A_l = \frac{1}{2\epsilon_0 R^{l-1}} \int_0^\pi \left[-\frac{32}{35}P_4(\cos\theta)P_l(\cos\theta)\sin\theta + \frac{22}{21}P_2(\cos\theta)P_l(\cos\theta)\sin\theta \right.$$
$$\left. -\frac{2}{15}P_0(\cos\theta)P_l(\cos\theta)\sin\theta \right]d\theta.$$

If $l = 4$, we have

$$A_4 = \frac{1}{2\epsilon_0 R^{4-1}} \int_0^\pi -\frac{32}{35}\left(\frac{35\cos^4\theta - 30\cos^2\theta + 3}{8}\right)^2 \sin\theta\, d\theta = -\frac{32}{315\epsilon_0 R^3}.$$

If $l = 2$, we have

$$A_2 = \frac{1}{2\epsilon_0 R^{2-1}} \int_0^\pi \frac{22}{21}\left(\frac{3\cos^2\theta - 1}{2}\right)^2 \sin\theta\, d\theta = \frac{22}{105\epsilon_0 R}.$$

If $l = 0$, we have

$$A_0 = \frac{1}{2\epsilon_0 R^{0-1}} \int_0^\pi -\frac{2}{15}\sin\theta\, d\theta = -\frac{2R}{15\epsilon_0}.$$

We can now find B_4, B_2, and B_0,

$$B_4 = A_4 R^{2(4)+1} = -\frac{32R^6}{315\epsilon_0}$$

$$B_2 = A_2 R^{2(2)+1} = \frac{22R^4}{105\epsilon_0}$$

$$B_0 = A_0 R^{2(0)+1} = -\frac{2R^2}{15\epsilon_0}.$$

Therefore, inside we have

$$V_{in}(r, \theta) = A_0 + A_2 r^2 P_2(\cos\theta) + A_4 r^4 P_4(\cos\theta).$$

So

$$V_{\text{in}}(r, \theta) = \frac{R}{\epsilon_0}\left[-\frac{2}{15} + \frac{22}{105\epsilon_0}\left(\frac{r}{R}\right)^2\left(\frac{3\cos^2\theta - 1}{2}\right)\right.$$

$$\left. - \frac{32}{315\epsilon_0}\left(\frac{r}{R}\right)^4\left(\frac{35\cos^4\theta - 30\cos^2\theta + 3}{8}\right)\right]$$

and outside we have

$$V_{\text{out}}(r, \theta) = \frac{B_0}{r^1} + \frac{B_2}{r^{2+1}}P_2(\cos\theta) + \frac{B_4}{r^{4+1}}P_4(\cos\theta).$$

So

$$V_{\text{out}}(r, \theta)$$

$$= \frac{R^2}{\epsilon_0 r}\left[-\frac{2}{15} + \frac{22}{105}\left(\frac{R}{r}\right)^2\left(\frac{3\cos^2\theta - 1}{2}\right) - \frac{32}{315}\left(\frac{R}{r}\right)^4\left(\frac{35\cos^4\theta - 30\cos^2\theta + 3}{8}\right)\right].$$

Problem 3.10. An infinitely long cylindrical shell of radius R is held at a potential $V_0(\phi) = \alpha\cos(4\phi)$. Find the potential inside and outside the shell.

Solution Our general solution in cylindrical coordinates is given by

$$V(s, \phi) = a_0 + b_0\ln(s) + \sum_{k=1}^{\infty}\left\{s^k\left[a_k\cos(k\phi) + b_k\sin(k\phi)\right]\right.$$

$$\left. + s^{-k}\left[c_k\cos(k\phi) + d_k\sin(k\phi)\right]\right\}.$$

Inside:
Here, we must have $b_0 = c_k = d_k = 0$, otherwise the potential would blow up at the center. Our potential becomes

$$V(s, \phi) = a_0 + \sum_{k=1}^{\infty}s^k\left[a_k\cos(k\phi) + b_k\sin(k\phi)\right].$$

At the surface, we have

$$V(R, \phi) = a_0 + \sum_{k=1}^{\infty}R^k\left[a_k\cos(k\phi) + b_k\sin(k\phi)\right] = \alpha\cos(4\phi).$$

Note we have $a_0 = 0$, $b_k = 0$, and $a_k = 0$, except for a_4. So

$$R^4 a_4\cos(4\phi) = \alpha\cos(4\phi) \rightarrow a_4 = \frac{\alpha}{R^4}.$$

Therefore, for $s \leqslant R$, we have

$$V(s, \phi) = \alpha\left(\frac{s}{R}\right)^4\cos(4\phi).$$

Outside:

Here, we must have $b_0 = a_k = b_k = 0$, otherwise the potential would blow up as $s \to \infty$. Also, since we must have $V \to 0$ as $s \to \infty$, $a_0 = 0$. Our potential becomes

$$V(s, \phi) = \sum_{k=1}^{\infty} s^{-k} \Big[c_k \cos(k\phi) + d_k \sin(k\phi) \Big].$$

At the surface, we have

$$V(R, \phi) = \sum_{k=1}^{\infty} R^{-k} \Big[c_k \cos(k\phi) + d_k \sin(k\phi) \Big] = \alpha \cos(4\phi).$$

Note we have $d_k = 0$ and $c_k = 0$, except for c_4. So

$$R^{-4} c_4 \cos(4\phi) = \alpha \cos 4\phi \to c_4 = \alpha R^4.$$

Therefore, for $s \geqslant R$, we have

$$V(s, \phi) = \alpha \left(\frac{R}{s} \right)^4 \cos(4\phi).$$

Problem 3.11. Given an infinitely long cylindrical shell of radius R and surface charge $\sigma_0(\phi) = \alpha \cos(2\phi) + \beta \sin(3\phi)$, find the potential inside and outside the cylinder.

Solution Our general solution in cylindrical coordinates is given by

$$V(s, \phi) = a_0 + b_0 \ln(s) + \sum_{k=1}^{\infty} \Big\{ s^k \Big[a_k \cos(k\phi) + b_k \sin(k\phi) \Big]$$

$$+ s^{-k} \Big[c_k \cos(k\phi) + d_k \sin(k\phi) \Big] \Big\}.$$

Inside, we must have $b_0 = c_k = d_k = 0$, otherwise the potential would blow up at the center. Our potential becomes

$$V_{\text{in}}(s, \phi) = a_0 + \sum_{k=1}^{\infty} s^k \Big[a_k \cos(k\phi) + b_k \sin(k\phi) \Big].$$

Outside we must have $b_0 = a_k = b_k = 0$, otherwise the potential would blow up as $s \to \infty$. Also, since we must have $V \to 0$ as $s \to \infty$, $a_0 = 0$. Our potential becomes

$$V_{\text{out}}(s, \phi) = \sum_{k=1}^{\infty} s^{-k} \Big[c_k \cos(k\phi) + d_k \sin(k\phi) \Big].$$

At the surface, we have

$$\left(\frac{\partial V_{\text{out}}}{\partial s} - \frac{\partial V_{\text{in}}}{\partial s} \right) \bigg|_{s=R} = -\frac{1}{\epsilon_0} \sigma_0(\phi),$$

where

$$\frac{\partial V_{\text{out}}}{\partial s} = \sum_{k=1}^{\infty} -ks^{-k-1}\Big[c_k \cos(k\phi) + d_k \sin(k\phi) \Big]$$

and

$$\frac{\partial V_{\text{in}}}{\partial s} = \sum_{k=1}^{\infty} ks^{k-1}\Big[a_k \cos(k\phi) + b_k \sin(k\phi) \Big].$$

Thus,

$$\left(\frac{\partial V_{\text{out}}}{\partial s} - \frac{\partial V_{\text{in}}}{\partial s}\right)\bigg|_{s=R} = \sum_{k=1}^{\infty} -kR^{-k-1}\Big[c_k \cos(k\phi) + d_k \sin(k\phi) \Big]$$

$$- kR^{k-1}\Big[a_k \cos(k\phi) + b_k \sin(k\phi) \Big]$$

$$= -\frac{\alpha \cos(2\phi) + \beta \sin(3\phi)}{\epsilon_0}.$$

From this, we can see that $c_k = a_k = 0$, except when $k = 2$, and $d_k = b_k = 0$, except when $k = 3$. This means

$$2\cos(2\phi)\Big(R^{-3}c_2 + Ra_2\Big) + 3\sin(3\phi)\Big(R^{-4}d_3 + R^2 b_3\Big) = \frac{\alpha \cos(2\phi)}{\epsilon_0} + \frac{\beta \sin(3\phi)}{\epsilon_0}.$$

Separating out the sine and cosine term, we have

$$2\Big(R^{-3}c_2 + Ra_2\Big) = \frac{\alpha}{\epsilon_0} \rightarrow c_2 = R^3\left(\frac{\alpha}{2\epsilon_0} - Ra_2\right)$$

and

$$3(R^{-4}d_3 + R^2 b_3) = \frac{\beta}{\epsilon_0} \rightarrow d_3 = R^4\left(\frac{\beta}{3\epsilon_0} - R^2 b_3\right).$$

Since V is continuous, we have

$$V_{\text{out}}(R, \phi) = V_{\text{in}}(R, \phi)$$

$$R^{-2}c_2 \cos(2\phi) + R^{-3}d_3 \sin(3\phi) = a_0 + R^2 a_2 \cos(2\phi) + R^3 b_3 \sin(3\phi).$$

We can see that $a_0 = 0$. Also, considering the sine and cosine terms separately, we have

$$R^{-2}c_2 = R^2 a_2.$$

Substitution of c_2 yields

$$R^{-2}R^3\left(\frac{\alpha}{2\epsilon_0} - Ra_2\right) = R^2 a_2$$

so

$$a_2 = \frac{\alpha}{4R\epsilon_0}.$$

Also

$$R^{-3}d_3 = R^3 b_3.$$

Substitution of d_3 yields

$$R^{-3}R^4\left(\frac{\beta}{3\epsilon_0} - R^2 b_3\right) = R^3 b_3$$

So

$$b_3 = \frac{\beta}{6R^2\epsilon_0}.$$

Therefore,

$$c_2 = R^3\left(\frac{\alpha}{2\epsilon_0} - R\frac{\alpha}{4R\epsilon_0}\right) = \frac{\alpha R^3}{4\epsilon_0}$$

and

$$d_3 = R^4\left(\frac{\beta}{3\epsilon_0} - R^2\frac{\beta}{6R^2\epsilon_0}\right) = \frac{\beta R^4}{6\epsilon_0}.$$

Combining everything, the potential inside is

$$V_{in}(s, \phi) = s^2\frac{\alpha}{4R\epsilon_0}\cos(2\phi) + s^3\frac{\beta}{6R^2\epsilon_0}\sin(3\phi)$$

$$= \frac{R}{\epsilon_0}\left[\frac{\alpha}{4}\left(\frac{s}{R}\right)^2\cos(2\phi) + \frac{\beta}{6}\left(\frac{s}{R}\right)^3\sin(3\phi)\right]$$

and outside is

$$V_{out}(s, \phi) = s^{-2}\frac{\alpha R^3}{4\epsilon_0}\cos(2\phi) + s^{-3}\frac{\beta R^4}{6\epsilon_0}\sin(3\phi)$$

$$= \frac{R}{\epsilon_0}\left[\frac{\alpha}{4}\left(\frac{R}{s}\right)^2\cos(2\phi) + \frac{\beta}{6}\left(\frac{R}{s}\right)^3\sin(3\phi)\right].$$

Problem 3.12. The electric potential varies as $\frac{1}{r}$ for a monopole, as $\frac{1}{r^2}$ for a dipole, as $\frac{1}{r^3}$ for a quadrupole, and as $\frac{1}{r^4}$ for an octopole. How will the electric potential depend on r for a mutipole with n charges (n being a k power of 2, $n = 2^k$)?

Solution

	Number of charges	Potential
Monopole	$n = 2^0 = 1; k = 0$	$V \sim \frac{1}{r^{k+1}} = \frac{1}{r}$
Dipole	$n = 2^1 = 2; k = 1$	$V \sim \frac{1}{r^{k+1}} = \frac{1}{r^2}$
Quadrupole	$n = 2^2 = 4; k = 2$	$V \sim \frac{1}{r^{k+1}} = \frac{1}{r^3}$
Octopole	$n = 2^3 = 8; k = 3$	$V \sim \frac{1}{r^{k+1}} = \frac{1}{r^4}$
Multipole	$n = 2^k; k$	$V \sim \frac{1}{r^{k+1}}$

Problem 3.13. Let us consider an electric dipole with charges q and $-q$ situated at distance d from each other, shown below. Calculate the electric potential at a point P in the far approximation $r \gg d$.

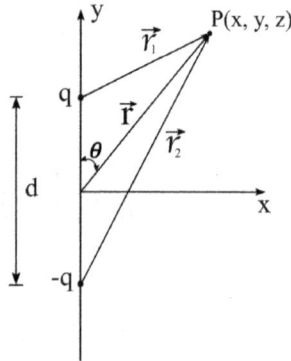

Solution The total electric potential is obtained by superposition

$$V(\vec{r}) = V_1 + V_2 = \frac{1}{4\pi\varepsilon_0}\left(\frac{q}{r_1} + \frac{-q}{r_2}\right).$$

From the law of cosines

$$r_1^2 = \left(\frac{d}{2}\right)^2 + r^2 - 2\frac{d}{2}r\cos\theta$$

and

$$r_2^2 = \left(\frac{d}{2}\right)^2 + r^2 - 2\frac{d}{2}r\cos(\pi - \theta).$$

Note that $\cos(\pi - \theta) = -\cos\theta$. We can rewrite our potential as

$$V = \frac{q}{4\pi\varepsilon_0}\left(\frac{1}{\sqrt{\frac{d^2}{4} + r^2 - rd\cos\theta}} - \frac{1}{\sqrt{\frac{d^2}{4} + r^2 + rd\cos\theta}}\right)$$

or

$$V = \frac{q}{4\pi\varepsilon_0}\left(\frac{1}{r\sqrt{1 + \frac{d^2}{4r^2} - \frac{d}{r}\cos\theta}} - \frac{1}{r\sqrt{1 + \frac{d^2}{4r^2} + \frac{d}{r}\cos\theta}}\right).$$

When $r \gg d$, $\frac{d^2}{4r^2}$ is very small and can be ignored. If we consider $x = \frac{d}{r}\cos\theta \ll 1$, we can use the binomial theorem and obtain

$$(1 + x)^{-\frac{1}{2}} \cong 1 - \frac{x}{2}$$

and

$$(1 - x)^{-\frac{1}{2}} \cong 1 + \frac{x}{2}.$$

From this, we have

$$\frac{1}{r\sqrt{1 - \frac{d}{r}\cos\theta}} = \frac{1}{r}\left(1 + \frac{d}{2r}\cos\theta\right)$$

and

$$\frac{1}{r\sqrt{1 + \frac{d}{r}\cos\theta}} = \frac{1}{r}\left(1 - \frac{d}{2r}\cos\theta\right).$$

Therefore,

$$V = \frac{q}{4\pi\varepsilon_0}\left[\frac{1}{r}\left(1 + \frac{d}{2r}\cos\theta\right) - \frac{1}{r}\left(1 - \frac{d}{2r}\cos\theta\right)\right]$$

$$= \frac{q}{4\pi\varepsilon_0 r}\left[1 + \frac{d}{2r}\cos\theta - 1 + \frac{d}{2r}\cos\theta\right]$$

$$V = \frac{qd}{4\pi\varepsilon_0 r^2}\cos\theta.$$

Taking the dipole moment $\vec{p} = q\vec{d}$, we have

$$V = \frac{\vec{p}\cdot\hat{r}}{4\pi\varepsilon_0 r^2}.$$

Problem 3.14. Find the electric field of the dipole in problem 3.13, centered at the origin with the dipole moment $\vec{p}$ in the z-direction.

Solution From problem 3.13, the electric potential is given by

$$V(\vec{r}) = \frac{qd}{4\pi\varepsilon_0 r^2}\cos\theta = \frac{\vec{p}\cdot\hat{r}}{4\pi\varepsilon_0 r^2} = \frac{p\cos\theta}{4\pi\varepsilon_0 r^2}.$$

We can find the field from the potential using

$$\vec{E} = -\nabla V.$$

We need to use the gradient in spherical coordinates

$$E_r = -\frac{\partial V}{\partial r} = \frac{2p\cos\theta}{4\pi\varepsilon_0 r^3}$$

$$E_\theta = -\frac{1}{r}\frac{\partial V}{\partial\theta} = \frac{p\sin\theta}{4\pi\varepsilon_0 r^3}$$

$$E_\phi = -\frac{1}{r\sin\theta}\frac{\partial V}{\partial\phi} = 0.$$

Therefore, the electric field due to the dipole is

$$\vec{E}_{\text{dipole}}(r,\theta) = \frac{p}{4\pi\varepsilon_0 r^3}\left(2\cos\theta\,\hat{r} + \sin\theta\,\hat{\theta}\right).$$

Problem 3.15. Two point charges $+4q$ and $-q$ are separated by a distance d. The first charge is placed at $(0, 0, d)$ and the second one at the origin. Find: (a) the monopole moment; (b) the dipole moment; (c) the electric potential in spherical coordinates for $r \gg d$. Include only the monopole and dipole contributions.

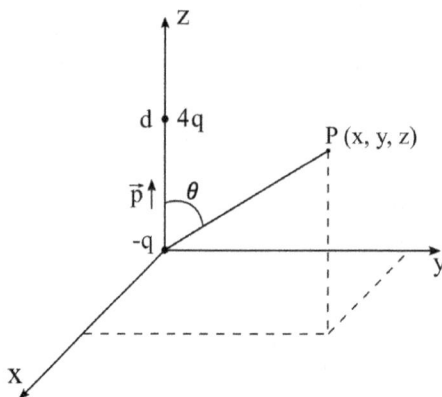

Solution

(a) Monopole moment:

$$Q = 4q - q = 3q.$$

(b) Dipole moment:

$$\vec{p} = \sum_{i=1}^{2} q_i \vec{r}_i = -q(0, 0, 0) + 4q(0, 0, d) = 4qd\hat{z}.$$

(c) The electric potential:

$$V(\vec{r}) = \frac{1}{4\pi\varepsilon_0}\left(\frac{1}{r}\sum_i q_i + \frac{1}{r^2}\sum_i q_i r_i' \cos\theta_i' + \cdots\right) = \frac{1}{4\pi\varepsilon_0}\left(\frac{3q}{r} + \frac{1}{r^2}\vec{p}\cdot\hat{r}\right)$$

$$= \frac{1}{4\pi\varepsilon_0}\left(\frac{3q}{r} + \frac{p\cos\theta}{r^2}\right) = \frac{1}{4\pi\varepsilon_0}\left(\frac{3q}{r} + \frac{4qd\cos\theta}{r^2}\right).$$

Bibliography

Byron F W and Fuller R W 1992 *Mathematics of Classical and Quantum Physics* (New York: Dover)

Griffiths D J 1999 *Introduction to Electrodynamics* 3rd edn (Englewood Cliffs, NJ: Prentice Hall)

Griffiths D J 2013 *Introduction to Electrodynamics* 4th edn (New York: Pearson)

Halliday D, Resnick R and Walker J 2010 *Fundamentals of Physics* 9th edn (New York: Wiley)

Halliday D, Resnick R and Walker J 2013 *Fundamentals of Physics* 10th edn (New York: Wiley)

Jackson J D 1998 *Classical Electrodynamics* 3rd edn (New York: Wiley)

Rogawski J 2011 *Calculus: Early Transcendentals* 2nd edn (San Francisco, CA: Freeman)

Electromagnetism
Problems and solutions
Carolina C Ilie and Zachariah S Schrecengost

Chapter 4

Magnetostatics

This chapter introduces magnetic fields in a vacuum and the methods for calculating the magnetic field. Magnetic fields are intrinsically determined by electric charges in motion. We imagine these small currents as magnetic dipoles. From the general Biot–Savart law, to the more straightforward Ampère's law applicable to configurations with higher degree of symmetry, the suggested problems constitute good practice in magnetostatics.

4.1 Theory

4.1.1 Magnetic force

A charge q moving with velocity $\vec{v}$ in a magnetic field $\vec{B}$ experiences a force given by

$$\vec{F}_{\mathrm{m}} = q\vec{v} \times \vec{B}.$$

4.1.2 Force on a current carrying wire

The force on a current carrying wire in a magnetic field $\vec{B}$ is

$$\vec{F}_{\mathrm{m}} = \int I\left(\mathrm{d}\vec{\ell} \times \vec{B}\right).$$

4.1.3 Volume current density

The current density of a current $\vec{I}$ is

$$\vec{J} = \frac{\mathrm{d}\vec{I}}{\mathrm{d}a_{\perp}}$$

and the current density of a charge density ρ moving at velocity $\vec{v}$ is

$$\vec{J} = \rho\vec{v}.$$

doi:10.1088/978-1-6817-4429-2ch4   4-1

4.1.4 Continuity equation

The divergence of the charge density $\vec{J}$ is related to the charge density ρ by

$$\nabla \cdot \vec{J} = -\frac{\partial \rho}{\partial t}.$$

4.1.5 Biot–Savart law

The magnetic field due to current distributions is given by

$$\vec{B}(\vec{r}) = \frac{\mu_0 I}{4\pi} \int \frac{\mathrm{d}\vec{\ell}' \times \hat{r}}{r^2}$$

$$\vec{B}(\vec{r}) = \frac{\mu_0}{4\pi} \int \frac{\vec{K}(\vec{r}') \times \hat{r}}{r^2} \mathrm{d}a'$$

$$\vec{B}(\vec{r}) = \frac{\mu_0}{4\pi} \int \frac{\vec{J}(\vec{r}') \times \hat{r}}{r^2} \mathrm{d}\tau'.$$

4.1.6 Divergence of $\vec{B}$

Given magnetic field $\vec{B}$, we have

$$\nabla \cdot \vec{B} = 0.$$

4.1.7 Ampère's law

Given magnetic field $\vec{B}$, we have

$$\nabla \times \vec{B} = \mu_0 \vec{J}.$$

By applying Stoke's law, we also have

$$\oint_{\mathcal{S}} \vec{B} \cdot \mathrm{d}\vec{\ell} = \mu_0 I_{\text{enc}},$$

where

$$I_{\text{enc}} = \int \vec{J} \cdot \mathrm{d}\vec{a}.$$

4.1.8 Vector potential

The vector potential due to current distributions is given by

$$\vec{A}(\vec{r}) = \frac{\mu_0 I}{4\pi} \int \frac{1}{r} \mathrm{d}\vec{\ell}'$$

$$\vec{A}(\vec{r}) = \frac{\mu_0}{4\pi} \int \frac{\vec{K}(\vec{r}')}{r} \mathrm{d}a'$$

$$\vec{A}(\vec{r}) = \frac{\mu_0}{4\pi} \int \frac{\vec{J}(\vec{r}')}{r} \mathrm{d}\tau'.$$

Also,

$$\vec{B} = \nabla \times \vec{A}$$

and

$$\nabla^2 \vec{A} = \mu_0 \vec{J}.$$

4.1.9 Magnetic dipole moment

The magnetic dipole moment due to a current I is

$$\vec{m} = \int I \, d\vec{a}.$$

4.1.10 Magnetic field due to dipole moment

Given magnetic dipole moment $\vec{m}$, the magnetic field is

$$\vec{B}_{\text{dip}} = \frac{\mu_0 m}{4\pi r^3}\left(2 \cos \theta \, \hat{r} + \sin \theta \, \hat{\theta}\right).$$

4.2 Problems and solutions

Problem 4.1. A proton travels through a uniform magnetic and electric field. The magnetic field is $\vec{B} = a\hat{y}$, where a is a positive constant. If at one moment the velocity of the proton is $\vec{v} = b\hat{z}$, where b is a positive constant, what is the force acting on the proton if the electric field is $\vec{E} = -c\hat{x}$?

Solution

$$\vec{F} = q\left(\vec{E} + \vec{v} \times \vec{B}\right) = q\left[-c\hat{x} + (b\hat{z}) \times (a\hat{y})\right] = q\left[-c\hat{x} + ab(-\hat{x})\right] = (c\hat{x} + ab\hat{x})$$

$$\vec{F} = -q(c + ab)\hat{x}.$$

Problem 4.2. A particle of charge q enters a region of uniform magnetic field $\vec{B}$ (out of the page, in the z-direction) with an initial velocity $\vec{v}$ (in the x-direction). The particle is deflected a distance y above the initial direction. If the region has a width of x, find the sign of the charge and the deflected distance y as a function of q, v, B, and x.

Solution

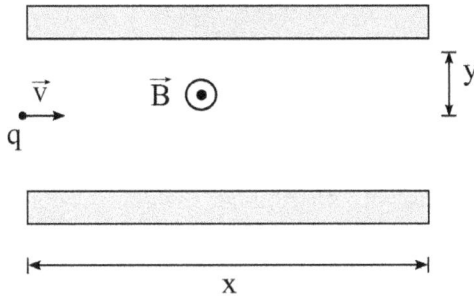

$$\vec{F} = q\vec{v} \times \vec{B}$$

Since the charge is deflected as shown, the charge is negative (determined from the right-hand rule). In the x-direction we have no force, and therefore no acceleration

$$x = vt.$$

In the y-direction

$$ma = |q|vB \sin 90.$$

So

$$a = \frac{|q|vB}{m}$$

and

$$y = y_0 + v_{oy}t + \frac{a_y t^2}{2} = \frac{a_y t^2}{2} = \frac{|q|vBt^2}{2m}.$$

By substituting the time

$$t = \frac{x}{v}$$

we obtain

$$y = \frac{|q|vB\frac{x^2}{v^2}}{2m} = \frac{|q|Bx^2}{2mv} = \frac{|q|Bx^2}{2p},$$

where p is the momentum of the particle.

Problem 4.3. The current density in a wire of circular cross section of radius R is dependent on the distance from the axis, given by $\vec{J} = ks^2\hat{z}$, where k is a constant. Find a) the total current in the wire and b) the current density if the current in a) is uniformly distributed.

Solution

a) Given current density $\vec{J} = ks^2\hat{z}$, the current is

$$I = \int \vec{J} \cdot d\vec{a} = \int \left(ks^2\hat{z}\right) \cdot \left(s\, d\phi\, ds\, \hat{z}\right) = \int_0^{2\pi} d\phi \int_0^R ks^3 ds = 2\pi k \frac{s^4}{4}\Big|_0^R = \frac{\pi k R^4}{2}.$$

b) If this current was uniformly distributed, the current density is simply

$$J = \frac{I}{\text{area}} = \frac{1}{\pi R^2}\frac{\pi k R^4}{2} = \frac{k R^2}{2}.$$

Problem 4.4.

a) In the famous experiment of J J Thompson, he measured the charge to mass $\frac{q}{m}$ ratio of the catode rays. Find $\frac{q}{m}$ when you know B, R, and v, and that $\vec{B}$ is perpendicular to $\vec{v}$.

b) He also had the beams going in a region with perpendicular electric field and magnetic field and 'tuned' them such that the electrons left the region with unchanged direction. If the speed of the electrons is v and the magnetic field is $\vec{B}$, what should be the value of the electric field?

Solution

a) The magnitude of the magnetic force is given by

$$\left|\vec{F}_m\right| = |q|\vec{v} \times \vec{B},$$

where we have $\vec{v} \perp \vec{B}$. Also

$$\vec{F}_m = \vec{F}_{\text{centripetal}}.$$

So,

$$F_m = |q|vB$$

and

$$F_{\text{cent}} = \frac{mv^2}{R}.$$

Therefore from

$$|q|vB = \frac{mv^2}{R}$$

we have

$$\frac{|q|}{m} = \frac{v}{BR}.$$

b) Setting the magnetic and electric forces as equal, we have

$$\vec{F}_{\mathrm{m}} = \vec{F}_{\mathrm{e}} \rightarrow q\vec{v} \times \vec{B} = q\vec{E}.$$

Dividing by q and expressing this in terms of magnitudes, we have

$$vB \sin 90 = E$$

so

$$E = vB.$$

Problem 4.5. Find the magnetic field at:
a) The center of a circular wire loop of radius R carrying current I.

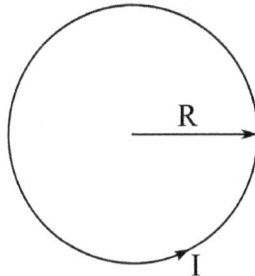

b) The center of a wire loop that consists of half a loop of radius R and half a square loop of side $2R$, carrying current I.

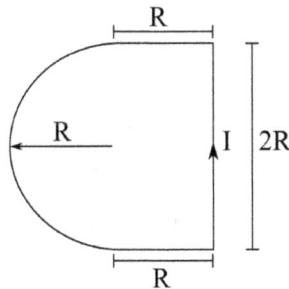

Solution
a) The Biot–Savart law states

$$\vec{B} = \frac{\mu_0 I}{4\pi} \int \frac{d\vec{\ell} \times \hat{r}}{r^2},$$

where

$$d\vec{\ell} = R \, d\vec{\phi} = R \, d\phi \, \hat{\phi}$$

and

$$\hat{r} = -\hat{r}.$$

Since $r = R$, the Biot–Savart law becomes

$$\vec{B} = \frac{\mu_0 I R}{4\pi R^2} \int\limits_{0}^{2\pi} \hat{\phi} \times \hat{r} \, d\phi = \frac{\mu_0 I}{2R} \hat{z} \quad \text{(out of page)}.$$

b) From part a), we can determine the field contribution due to the circular part is

$$\vec{B}_c = \frac{\mu_0 I}{4R} \hat{z},$$

which is half that of the full loop. As for the square, we consider the field R above the wire. We have

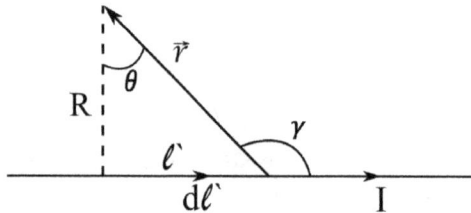

So

$$\vec{B} = \frac{\mu_0 I}{4\pi} \int \frac{d\vec{\ell}' \times \hat{r}}{r^2},$$

where $d\vec{\ell}' \times \hat{r}$ points in the $\hat{z}$-direction (out of the page). Also,

$$d\ell' \sin \gamma = d\ell' \cos \theta$$

and

$$\ell' = R \tan \theta \rightarrow d\ell' = \frac{R}{\cos^2 \theta} d\theta$$

and

$$r^2 = \ell'^2 + R^2 \rightarrow \frac{1}{r^2} = \frac{\cos^2\theta}{R^2}.$$

Therefore,

$$\vec{B} = \frac{\mu_0 I}{4\pi} \int_{\theta_1}^{\theta_2} \left(\frac{\cos^2\theta}{R^2}\right)\left(\frac{R}{\cos^2\theta}\right)\cos\theta\,d\theta = \frac{\mu_0 I}{4\pi R}(\sin\theta_2 - \sin\theta_1).$$

So for each R-lengthed segment (i) and (ii), we have $\theta_1 = 0$ and $\theta_2 = \frac{\pi}{4}$, and for the $2R$-lengthed segment, we have $\theta_1 = -\frac{\pi}{4}$ and $\theta_2 = \frac{\pi}{4}$. So

$$\vec{B} = \vec{B}_c + 2\left(\frac{\mu_0 I}{4\pi R}\sin\frac{\pi}{4}\hat{z}\right) + \frac{\mu_0 I}{4\pi R}\left[\sin\left(\frac{\pi}{4}\right) - \sin\left(-\frac{\pi}{4}\right)\right]\hat{z}$$

$$= \frac{\mu_0 I}{4\pi R}\left(\pi + 2\sqrt{2}\right)\hat{z} \qquad \text{(out of page)}.$$

Problem 4.6. Consider a cylindrical shell of radius R and length L, carrying σ and rotating at ω. Find the magnetic field d from the end of the shell (on the axis).

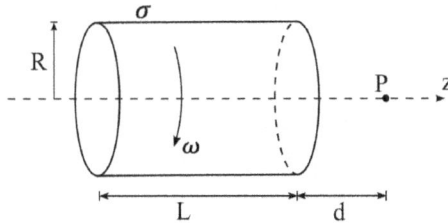

Solution Here we have

$$\vec{B} = \frac{\mu_0}{4\pi}\int\frac{\vec{K}\times\hat{r}}{r^2}da,$$

where

$$da = R\,dz\,d\phi$$

and $0 \leqslant z \leqslant L$. The surface charge is given by

$$\vec{K} = \sigma\vec{v} = \sigma\omega R\hat{\phi}.$$

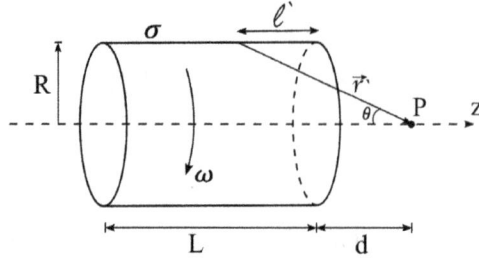

From the figure above, we have $\ell' = L - z$ and

$$r = \sqrt{(\ell' + d)^2 + R^2} = \sqrt{(L - z + d)^2 + R^2}.$$

Note that the field cancels such that the $\hat{z}$-component is the only component that survives. So

$$\left[\vec{K} \times \hat{r}\right]_z = \sigma \omega R \sin \theta \, \hat{z} = \sigma \omega R \frac{R}{r} \hat{z} = \frac{\sigma \omega R^2}{r} \hat{z}.$$

Putting everything together, we have

$$\vec{B} = \frac{\mu_0}{4\pi} \int_0^{2\pi} \int_0^L \frac{\sigma \omega R^2 R}{\left[(L - z + d)^2 + R^2\right]^{3/2}} \hat{z} \, \mathrm{d}z \, \mathrm{d}\phi$$

$$= \frac{\mu_0 \sigma \omega R^3 \hat{z}}{2} \int_0^L \frac{\mathrm{d}z}{\left[(L - z + d)^2 + R^2\right]^{3/2}}.$$

Therefore,

$$\vec{B} = \frac{\mu_0 \sigma \omega R}{2} \left[\frac{d + L}{\sqrt{R^2 + (d + L)^2}} - \frac{d}{\sqrt{R^2 + d^2}}\right] \hat{z}.$$

Problem 4.7. A hemisphere of radius R and charge density ρ is rotating at ω. Find the magnetic field d above the center.

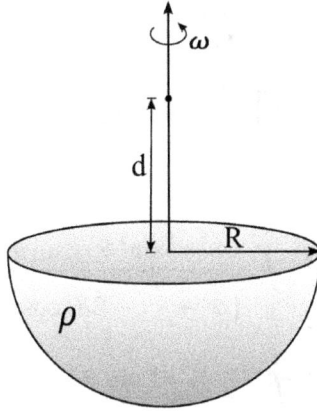

Solution Here we have

$$\vec{B} = \frac{\mu_0}{4\pi} \int \frac{\vec{J} \times \hat{r}}{r^2} d\tau,$$

where

$$d\tau = r^2 \sin\theta \, dr \, d\phi \, d\theta.$$

From the figure below, we can see that

$$r^2 = d^2 + r^2 - 2dr \cos\theta.$$

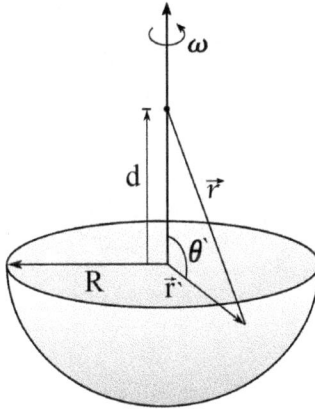

Also,

$$\vec{J} = \rho\vec{v} = \rho\omega r \sin\theta \, \hat{\phi}.$$

Note that the field cancels such that the $\hat{z}$-component is the only component that survives. So

$$\left[\vec{J} \times \hat{r}\right]_z = \rho \omega r \sin \theta \left(\frac{r \sin \theta}{r}\right)\hat{z}.$$

Therefore

$$\vec{B} = \frac{\mu_0}{4\pi} \int\limits_0^{2\pi} \int\limits_{\frac{\pi}{2}}^{\pi} \int\limits_0^R \frac{\rho \omega r^2 \sin^2 \theta \, r^2 \sin \theta \, \hat{z}}{\left(d^2 + r^2 - 2dr \cos \theta\right)^{\frac{3}{2}}} dr \, d\theta \, d\phi$$

$$= \frac{\mu_0 \rho \omega}{2} \int\limits_{\frac{\pi}{2}}^{\pi} \int\limits_0^R \frac{r^4 \sin^3 \theta \, \hat{z}}{\left(d^2 + r^2 - 2dr \cos \theta\right)^{\frac{3}{2}}} dr \, d\theta$$

$$\vec{B} = \frac{\mu_0 \rho \omega}{30d^3}\left[\sqrt{R^2 + d^2}\left(-2R^4 + d^2R^2 - 12d^4\right) + 2R^5 + 5d^3R^2 + 12d^5\right]\hat{z}.$$

Problem 4.8. A spherical shell of radius R, carrying σ and rotating at ω, is centered at the origin. Find the velocity a loop of wire, carrying λ with radius a centered at the origin, required to cancel the magnetic field at the center.

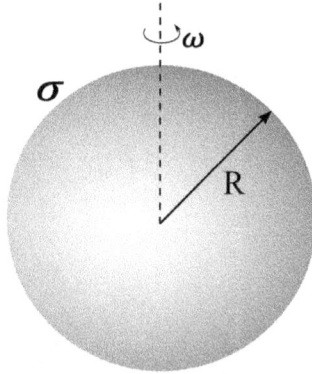

Solution First, we have

$$\vec{B} = \frac{\mu_0}{4\pi} \int \frac{\vec{K} \times \hat{r}}{r^2} da,$$

where

$$r = R$$

$$\hat{\imath} = -\hat{r}$$

$$\vec{K} = \sigma\vec{v} = \sigma\omega R \sin\theta \,\hat{\phi}$$

$$da = R^2 \sin\theta \,d\theta \,d\phi$$

and

$$\vec{K} \times \hat{\imath} = \sigma\omega R \sin^2\theta \,\hat{z}.$$

Putting this together, we have

$$\vec{B}_s = \frac{\mu_0}{4\pi} \int_0^\pi \int_0^{2\pi} \frac{\sigma\omega R \sin^2\theta \, R^2 \sin\theta \,\hat{z}}{R^2} d\phi \,d\theta = \frac{\mu_0\sigma\omega R \,\hat{z}}{2} \int_0^\pi \sin^3\theta \,d\theta$$

$$\vec{B}_s = \frac{2\mu_0\sigma\omega R}{3}\hat{z}.$$

Now a line of charge λ rotating at $\vec{v}$ 'looks' like a wire carrying current $\vec{I} = \lambda\vec{v} = \lambda\omega_l a\hat{\phi}$. From problem 4.5(a), we know this produces magnetic field

$$\vec{B}_l = \frac{\mu_0\lambda\omega_l a}{2a}\hat{z} = \frac{\mu_0\lambda\omega_l}{2}\hat{z}.$$

We want $\vec{B}_l + \vec{B}_s = 0$, so

$$\frac{\mu_0\lambda\omega_l}{2} + \frac{2\mu_0\sigma\omega R}{3} = 0 \rightarrow \frac{\lambda\omega_l}{2} = -\frac{2\sigma\omega R}{3}.$$

Therefore,

$$\omega_l = -\frac{4\omega\sigma R}{3\lambda}.$$

Problem 4.9. A long straight wire carries a steady current I. Obtain the magnetic field at a distance s from the wire.

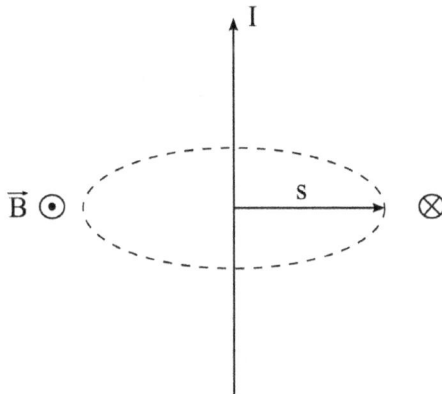

Solution Here, we apply Ampère's law, with an Amperian loop that is a circle centered on the wire and in a plane perpendicular to the wire. Ampère's law is

$$\oint_S \vec{B} \cdot d\vec{\ell} = \mu_0 I_{\text{enc}}.$$

Noticing $\vec{B} \parallel d\vec{\ell}$, it follows that B is a constant at a certain distance s from the wire. So the left-hand side is given by

$$\oint_S \vec{B} \cdot d\vec{\ell} = \oint_S B\, d\ell = B \oint_S d\ell = B2\pi s.$$

The enclosed current is just simply given by

$$I_{\text{enc}} = I$$

So

$$B = \frac{\mu_0 I}{2\pi s}.$$

Since the magnetic field is tangent on the circle at every point

$$\vec{B} = \frac{\mu_0 I}{2\pi s}\hat{\phi}.$$

Problem 4.10. An electric current flows through a long cylinder wire of radius a. Find the magnetic field inside and outside the wire, and plot it, in the following cases, where k is a constant with the appropriate units:
a) $I =$ constant (steady current).
b) Current density J is proportional to the distance from the axis: $J = ks$.
c) $J = ks^2$.

Solution
a) Here we have constant I. For $s > a$, our Amperian loop is given by

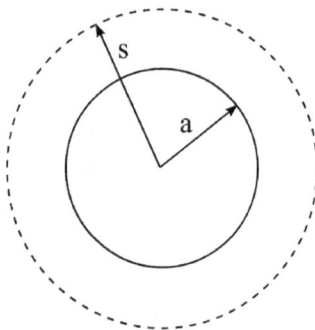

Ampère's law states

$$\oint_S \vec{B} \cdot \mathrm{d}\vec{\ell} = \mu_0 I_{\text{enc}}$$

In all cases, the left-hand side yields

$$\oint_S \vec{B} \cdot \mathrm{d}\vec{\ell} = \oint_S B \, \mathrm{d}\ell = B \oint_S \mathrm{d}\ell = B2\pi s.$$

Here we simply have

$$I_{\text{enc}} = I$$

So

$$B2\pi s = \mu_0 I \rightarrow B = \frac{\mu_0 I}{2\pi s}.$$

Applying the right-hand rule to a current coming out of the page, we have

$$\vec{B} = \frac{\mu_0 I}{2\pi s} \hat{\phi}.$$

For $s < a$, our Amperian loop is inside the wire, at radius s.

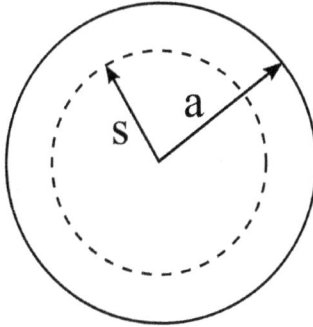

Again we have

$$\oint_S \vec{B} \cdot \mathrm{d}\vec{\ell} = \mu_0 I_{\text{enc}}.$$

Since I is uniform, the current density is constant,

$$J = \frac{I}{\pi a^2} = \frac{I_{\text{enc}}}{\pi s^2}$$

so

$$I_{\text{enc}} = I \frac{s^2}{a^2}.$$

Therefore

$$B = \frac{\mu_0 sI}{2\pi a^2} \rightarrow \vec{B} = \frac{\mu_0 sI}{2\pi a^2}\hat{\phi}.$$

The plot of the magnetic field is given below.

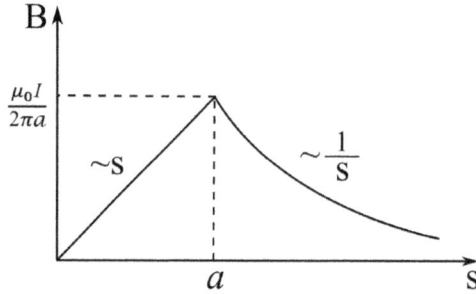

b) Here we have $J = ks$. For $s > a$, the Amperian loop is the same as part a) for $s > a$. We still have the left hand side of Ampère's law given by

$$\oint_S \vec{B} \cdot d\vec{\ell} = B2\pi s.$$

Now, since J is not constant, we need to integrate in order to find the enclosed current, so

$$I_{enc} = \int J \, da = \int_0^a Js \, ds \int_0^{2\pi} d\phi = \int_0^a ks^2 \, ds \int_0^{2\pi} d\phi = 2\pi \frac{ks^3}{3}\bigg|_0^a = \frac{2\pi ka^3}{3}.$$

Therefore,

$$B2\pi s = \frac{\mu_0 2\pi ka^3}{3} \rightarrow B = \frac{\mu_0 ka^3}{3s}$$

with

$$\vec{B} = \frac{\mu_0 ka^3}{3s}\hat{\phi}.$$

For $s < a$, we again have the same Amperian loop as part a). Here, our enclosed current is given by

$$I_{enc} = \int J \, da = \int_0^s ks' s' \, ds' \int_0^{2\pi} d\phi = 2\pi \frac{ks'^3}{3}\bigg|_0^s = \frac{2\pi ks^3}{3}.$$

Therefore,

$$B2\pi s = \frac{\mu_0 2\pi k s^3}{3} \rightarrow B = \frac{\mu_0 k s^2}{3}$$

with

$$\vec{B} = \frac{\mu_0 k s^2}{3}\hat{\phi}.$$

The plot of the magnetic field is given below.

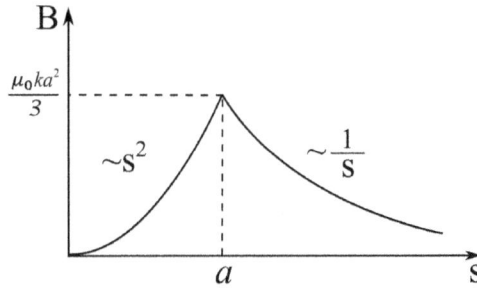

c) Now we have $J = ks^2$. For $s > a$, the enclosed current is given by

$$I_{enc} = \int J \, da = \int_0^a Js \, ds \int_0^{2\pi} d\phi = 2\pi \int_0^a ks^2 s \, ds = 2\pi \frac{ks^4}{4}\Big|_0^a = 2\pi \frac{ka^4}{4}.$$

Therefore,

$$B2\pi s = \mu_0 2\pi \frac{ka^4}{4} \rightarrow B = \frac{\mu_0 ka^4}{4s}$$

with

$$\vec{B} = \frac{\mu_0 ka^4}{4s}\hat{\phi}.$$

For $s < a$, we have

$$I_{enc} = \int J \, da = \int_0^s Js' \, ds' \int_0^{2\pi} d\phi = 2\pi \int_0^s ks'^2 s' \, ds'' = \frac{2\pi ks'^4}{4}\Big|_0^s = \frac{2\pi ks^4}{4}.$$

Therefore,

$$B = \frac{\mu_0 ks^3}{4}$$

with

$$\vec{B} = \frac{\mu_0 k s^3}{4}\hat{\phi}.$$

The plot of magnetic field is given below.

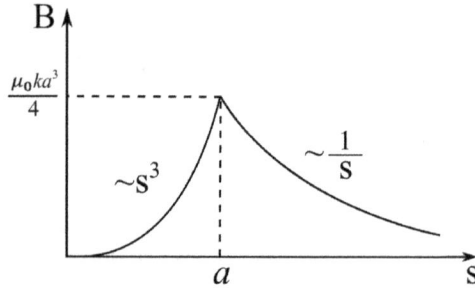

Problem 4.11. Use Ampère's law to obtain the magnetic field inside and outside a solenoid of $n = \frac{N}{L}$, where N is the number of turns, and L is the length of the solenoid. The solenoid is carrying the current I.

Solution Let us choose two Amperian loops given by

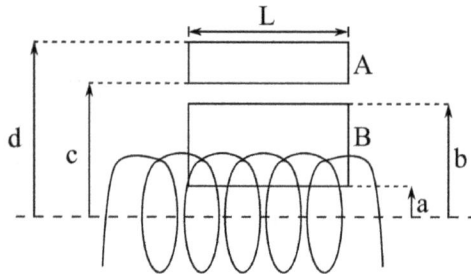

Starting with the outside loop (loop A), the magnetic field does not have any radial B_r component or B_ϕ. Ampère's law is given by

$$\oint_S \vec{B} \cdot d\vec{\ell} = \mu_0 I_{\text{enc}}.$$

Since we have $I_{enc} = 0$, $B(c) = B(d)$, but since $B \rightarrow 0$ for large distances, $B = 0$ outside the solenoid. For loop B, the left hand side of Ampère's law is given by

$$\oint_S \vec{B} \cdot d\vec{\ell} = \oint_S B \, d\ell = BL.$$

The sides perpendicular on the solenoid yield zero dot product, as the magnetic field is oriented parallel to the solenoid's axis in the z-direction (by the right-hand rule). The enclosed current is given by

$$I_{enc} = InL = IN.$$

Substituting these quantities into Ampère's law yields

$$BL = \mu_0 InL \rightarrow B = \mu_0 In = \frac{\mu_0 IN}{L}$$

with

$$\vec{B} = \mu_0 In\hat{z} = \frac{\mu_0 IN}{L}\hat{z}.$$

Problem 4.12. A current carrying empty cylinder of inner radius a and outer radius b has a current density J, which is proportional to the distance from the axis; $J = ks$, k constant. Find the magnetic field in all regions.

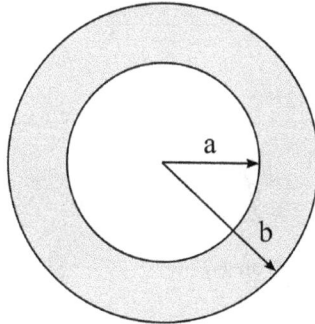

Solution There are three significant regions: $s < a$, $a < s < b$, and $s > b$. The easiest to find is the field for $s < a$,

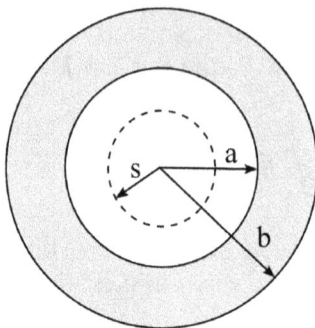

where the enclosed current is zero. Therefore,

$$\oint_S \vec{B} \cdot d\vec{\ell} = 0,$$

so $\vec{B} = 0$. For $a < s < b$, we have

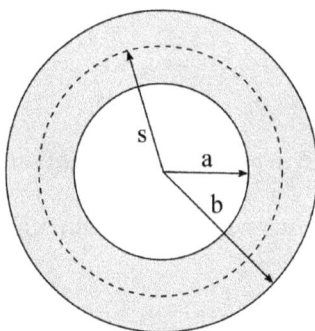

Using Ampère's law,

$$\oint_S \vec{B} \cdot d\vec{\ell} = \mu_0 I_{\text{enc}}.$$

We have the left-hand side is given by

$$\oint_S \vec{B} \cdot d\vec{\ell} = B2\pi s,$$

with enclosed current given by

$$I_{\text{enc}} = \int \vec{J} \cdot d\vec{a} = \int_a^s Js' \, ds' \int_0^{2\pi} d\phi = 2\pi \int_a^s ks' \, s' \, ds' = \frac{2\pi ks'^3}{3}\bigg|_a^s = \frac{2\pi k\left(s^3 - a^3\right)}{3}$$

$$B = \frac{\mu_o k\left(s^3 - a^3\right)}{3s}$$

with

$$\vec{B} = \frac{\mu_o k\left(s^3 - a^3\right)}{3s}\hat{\phi}.$$

For $s > b$, we have

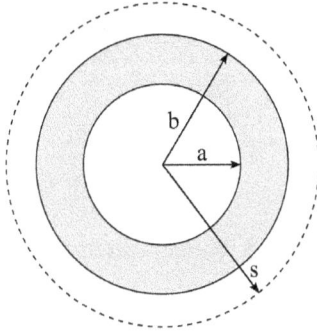

Again,

$$\oint_S \vec{B}\cdot \mathrm{d}\vec{\ell} = B2\pi s$$

with

$$I_{\mathrm{enc}} = \int \vec{J}\cdot \mathrm{d}\vec{a} = \int_a^b Js\,\mathrm{d}s\int_0^{2\pi} \mathrm{d}\phi = 2\pi\int_a^b ks\,s\,\mathrm{d}s = 2\pi\frac{ks^3}{3}\Big|_a^b = \frac{2\pi k\left(b^3 - a^3\right)}{3}.$$

Therefore,

$$B = \frac{k\left(b^3 - a^3\right)}{3s}$$

with

$$\vec{B} = \frac{k\left(b^3 - a^3\right)}{3s}\hat{\phi}.$$

Problem 4.13. Find the vector potential d above a spinning disk of radius R, with angular velocity ω and carrying σ.

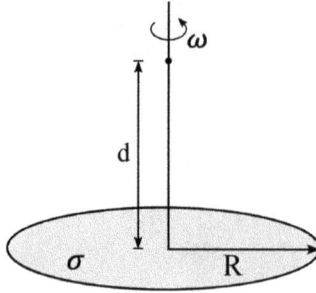

Solution We have

$$\vec{A} = \frac{\mu_0}{4\pi} \int \frac{\vec{K}}{r} da,$$

where

$$\vec{K} = \sigma \vec{v} = \sigma \omega r \hat{\phi}.$$

From

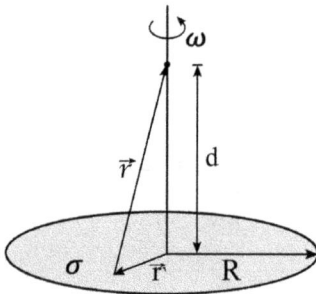

we also have

$$r = \sqrt{r^2 + d^2}$$

and

$$da = 2\pi r \, dr.$$

So

$$\vec{A} = \frac{\mu_0}{4\pi} \int_0^R \frac{\sigma \omega r 2\pi r \hat{\phi}}{\sqrt{r^2 + d^2}} dr = \frac{\mu_0 \sigma \omega \hat{\phi}}{2} \int_0^R \frac{r^2}{\sqrt{r^2 + d^2}} dr$$

$$= \frac{\mu_0 \sigma \omega}{4} \left[R\sqrt{R^2 + d^2} - d^2 \ln\left(R + \sqrt{R^2 + d^2}\right) + d^2 \ln d \right] \hat{\phi}$$

$$\vec{A} = \frac{\mu_0 \sigma \omega}{4} \left[R\sqrt{R^2 + d^2} + \ln\left(\frac{d}{R + \sqrt{R^2 + d^2}} \right) \right] \hat{\phi}.$$

Problem 4.14. What current density produces vector potential $\vec{A} = \sin \phi \, \hat{z}$?

Solution First, check $\nabla \cdot \vec{A} = 0$:

$$\nabla \cdot \vec{A} = \nabla \cdot (\sin \phi \, \hat{z}) = \frac{\partial}{\partial z}(\sin \phi) = 0.$$

Now, $\vec{B} = \nabla \times \vec{A}$ and $\nabla \times \vec{B} = \mu_0 \vec{J}$, so

$$\vec{B} = \nabla \times \vec{A} = \nabla \times (\sin \phi \, \hat{z}) = \frac{1}{s}\left[\frac{\partial}{\partial \phi}(\sin \phi) \right] \hat{s} = \frac{\cos \phi}{s} \hat{s}.$$

Also,

$$\mu_0 \vec{J} = \nabla \times \vec{B} = \nabla \times \left(\frac{\cos \phi}{s} \hat{s} \right) = \frac{1}{s}\left[-\frac{\partial}{\partial \phi}\left(\frac{\cos \phi}{s} \right) \right] \hat{z} = \frac{\sin \phi}{s^2} \hat{z}.$$

Therefore,

$$\vec{J} = \frac{\sin \phi}{\mu_0 s^2} \hat{z}.$$

This can be checked using the product rule

$$\nabla \times \left(\nabla \times \vec{A} \right) = \mu_0 \vec{J} = \nabla\left(\nabla \cdot \vec{A} \right) - \nabla^2 \vec{A}.$$

Since $\nabla \cdot \vec{A} = 0$, $\nabla^2 \vec{A} = \mu_0 \vec{J}$ where

$$-\nabla^2 \vec{A} = -\frac{1}{s}\frac{\partial^2}{\partial \phi^2}(\sin \phi)\hat{z} = -\frac{1}{s^2}(-\sin \phi)\hat{z} = \frac{\sin \phi}{s^2} \hat{z}.$$

So

$$\vec{J} = \frac{1}{\mu_0}\left(-\nabla^2 \vec{A} \right) = \frac{\sin \phi}{\mu_0 s^2} \hat{z}$$

as expected.

Problem 4.15. Find the vector potential inside and outside a wire of radius R that is carrying current density $\vec{J} = ks\hat{z}$, where k is a constant.

Solution We can find the field inside by

$$\oint_S \vec{B} \cdot d\vec{\ell} = \mu_0 I_{\text{enc}},$$

where

$$\oint_S \vec{B} \cdot d\vec{\ell} = B 2\pi s$$

and

$$I_{\text{enc}} = \int J \, da = \int_0^s ks' 2\pi s' \, ds' = \frac{2\pi k s^3}{3}.$$

So,

$$\oint_S \vec{B} \cdot d\vec{\ell} = B 2\pi s = \mu_0 \frac{2\pi k s^3}{3}$$

and

$$\vec{B} = \frac{\mu_0 k s^2}{3} \hat{\phi}.$$

We must have that $\vec{A}$ depends only on s and is in the direction of the current. So $\vec{A} = A(s)\hat{z}$ and $\nabla \times \vec{A} = \vec{B}$. Note

$$\nabla \times \vec{A} = -\frac{\partial A}{\partial s} \hat{\phi} = \vec{B} = \frac{\mu_0 k s^2}{3} \hat{\phi}.$$

Therefore,

$$dA = -\frac{\mu_0 k}{3} s^2$$

and

$$\vec{A} = \left(-\frac{\mu_0 k s^3}{9} + C \right) \hat{z}.$$

We will express this as

$$\vec{A} = -\frac{\mu_0 k}{9} \left(s^3 - \alpha^3 \right) \hat{z}.$$

Outside, our total current is

$$I_{\text{tot}} = I_{\text{enc}} = \int J \, da = \frac{2\pi k R^3}{3}.$$

From

$$\oint_S \vec{B} \cdot \mathrm{d}\vec{\ell} = \mu_0 I_{\text{enc}}$$

we have

$$B 2\pi s = \mu_0 \frac{2\pi k R^3}{3}$$

and

$$\vec{B} = \frac{\mu_0 k R^3}{3s} \hat{\phi}.$$

Again,

$$\mathrm{d}A = -B \, \mathrm{d}s = -\frac{\mu_0 k R^3}{3s} \mathrm{d}s$$

so

$$\vec{A} = \left(-\frac{\mu_0 k R^3}{3} \ln s + C \right) \hat{z}.$$

We will express this as

$$\vec{A} = -\frac{\mu_0 k R^3}{3} \left(\ln \frac{s}{\beta} \right) \hat{z}.$$

Since $\vec{A}$ is continuous at R,

$$-\frac{\mu_0 k}{9} \left(s^3 - \alpha^3 \right) = -\frac{\mu_0 k R^3}{3} \left(\ln \frac{s}{\beta} \right)$$

we have

$$R^3 - \alpha^3 = 3 R^3 \ln \frac{R}{\beta}$$

with

$$R^3 \left(1 - 3 \ln \frac{R}{\beta} \right) = \alpha^3$$

and

$$1 - 3 \ln \frac{R}{\beta} = \frac{\alpha^3}{R^3}.$$

If $\alpha = \beta = R$,

$$1 - 3\ln\left(\frac{R}{R}\right) = \frac{R^3}{R^3} \to 1 = 1.$$

So

$$\vec{A} = \begin{cases} -\dfrac{\mu_0 k}{9}\left(s^3 - R^3\right)\hat{z} & s < R \\[3mm] -\dfrac{\mu_0 k R^3}{3}\ln\dfrac{s}{R}\,\hat{z} & s > R \end{cases}.$$

Problem 4.16. A disk of radius R is carrying surface charge $\sigma = kr$, where k is a constant, and spinning at angular velocity ω. Find the magnetic dipole moment and the field it produces.

Solution We have

$$\vec{K} = \sigma\vec{v} = \sigma\omega r\hat{\phi} = \omega k r^2\hat{\phi}.$$

So

$$dI = \omega k r^2 dr \to I = \frac{\omega k r^3}{3}.$$

Therefore,

$$\vec{m} = \int I\, d\vec{a} = \frac{\omega k}{3}\int_0^R r^3 2\pi r\, dr\, \hat{z} = \frac{2\pi\omega k}{3}\int_0^R r^4\, dr$$

$$\vec{m} = \frac{2\pi\omega k R^5}{15}\hat{z}.$$

We have, in spherical coordinates,

$$\vec{B}_{\text{dip}} = \frac{\mu_0 m}{4\pi r^3}\left(2\cos\theta\,\hat{r} + \sin\theta\,\hat{\theta}\right),$$

which can be expressed in cylindrical coordinates by considering $r = \sqrt{s^2 + z^2}$, $\cos\theta = \frac{z}{r}$, $\sin\theta = \frac{s}{r}$, $\hat{\theta} = \frac{z}{r}\hat{s} - \frac{s}{r}\hat{z}$, and $\hat{r} = \frac{s}{r}\hat{s} + \frac{z}{r}\hat{z}$. Therefore,

$$\vec{B}_{\text{dip}} = \frac{\mu_0 m}{4\pi r^3}\left[2\left(\frac{z}{r}\right)\left(\frac{s}{r}\hat{s} + \frac{z}{r}\hat{z}\right) + \left(\frac{s}{r}\right)\left(\frac{z}{r}\hat{s} - \frac{s}{r}\hat{z}\right)\right]$$

$$= \frac{\mu_0 m}{4\pi r^5}\left(2zs\hat{s} + 2z^2\hat{z} + sz\hat{s} - s^2\hat{z}\right)$$

$$= \frac{\mu_0 m}{4\pi r^5}\left[3zs\hat{s} + \left(2z^2 - s^2\right)\hat{z}\right]$$

$$\vec{B}_{\text{dip}} = \frac{\mu_0 m}{4\pi} \frac{1}{\left(s^2 + z^2\right)^{5/2}}\left[3zs\hat{s} + \left(2z^2 - s^2\right)\hat{z}\right].$$

Substitution of m yields,

$$\vec{B}_{\text{dip}} = \frac{\mu_0 \omega k R^5}{30} \frac{1}{\left(s^2 + z^2\right)^{5/2}}\left[3zs\hat{s} + \left(2z^2 - s^2\right)\hat{z}\right].$$

Bibliography

Griffiths D J 1999 *Introduction to Electrodynamics* 3rd edn (Englewood Cliffs, NJ: Prentice Hall)

Griffiths D J 2013 *Introduction to Electrodynamics* 4th edn (New York: Pearson)

Halliday D, Resnick R and Walker J 2010 *Fundamentals of Physics* 9th edn (New York: Wiley)

Halliday D, Resnick R and Walker J 2013 *Fundamentals of Physics* 10th edn (New York: Wiley)

IOP Concise Physics

Electromagnetism
Problems and solutions
Carolina C Ilie and Zachariah S Schrecengost

Chapter 5

Electric fields in matter

Now we will address problems that deal with electric fields in matter, looking at problems involving dipole moments, media polarization, and electric displacement. Ideas developed in chapters 2 and 3 will be revisited and expanded upon in this chapter. Gauss's law is reformulated for electric displacement and various ways to calculate the energy of a configuration. Some of the techniques practiced in chapter 3 will be applied now, including the Laplace equation and Legendre polynomials.

5.1 Theory

5.1.1 Induced dipole moment of an atom in an electric field

Given an atom with polarizability α in an electric field $\vec{E}$, the induced dipole moment is

$$\vec{p} = \alpha\vec{E}.$$

5.1.2 Torque on a dipole due to an electric field

Given a dipole moment $\vec{p}$ in an electric field $\vec{E}$, the torque on the dipole is

$$\vec{N} = \vec{p} \times \vec{E}.$$

5.1.3 Force on a dipole

Given a dipole moment $\vec{p}$ in an electric field $\vec{E}$, the force on the dipole is

$$\vec{F} = \left(\vec{p} \cdot \nabla\right)\vec{E}.$$

5.1.4 Energy of a dipole in an electric field

Given a dipole moment $\vec{p}$ in an electric field $\vec{E}$, the energy of the dipole is

$$U = -\vec{p} \cdot \vec{E}.$$

5.1.5 Surface bound charge due to polarization $\vec{P}$

Given polarization $\vec{P}$ and normal vector $\hat{n}$, the surface bound charge is

$$\sigma_b = \vec{P} \cdot \hat{n}.$$

5.1.6 Volume bound charge due to polarization $\vec{P}$

Given polarization $\vec{P}$, the volume bound charge is

$$\rho_b = -\nabla \cdot \vec{P}.$$

5.1.7 Potential due to polarization $\vec{P}$

Given a volume $\mathcal{V}$, the potential due to polarization $\vec{P}(\vec{r})$ is

$$V(\vec{r}) = \frac{1}{4\pi\varepsilon_0} \int_{\mathcal{V}} \frac{\hat{r} \cdot \vec{P}(\vec{r}')}{r^2} \mathrm{d}\tau'.$$

5.1.8. Electric displacement

Given polarization $\vec{P}$ and electric field $\vec{E}$, the electric displacement is given by

$$\vec{D} = \varepsilon_0 \vec{E} + \vec{P}.$$

5.1.9 Gauss's law for electric displacement

Considering electric displacement $\vec{D}$ and free charge density ρ_{f}, Gauss's law can be written in differential form as

$$\nabla \cdot \vec{D} = \rho_{\mathrm{f}}$$

and in integral form as

$$\oint_S \vec{D} \cdot \mathrm{d}\vec{a} = q_{f_{\mathrm{enc}}},$$

where $q_{f_{\mathrm{enc}}}$ is the total free charge enclosed in the volume.

5.1.10 Linear dielectrics

Given a medium with electric susceptibility χ_e, the polarization is given by

$$\vec{P} = \varepsilon_0 \chi_e \vec{E},$$

where $\vec{E}$ is the total electric field. The electric displacement is now

$$\vec{D} = \varepsilon_0\vec{E} + \vec{P} = \left(\varepsilon_0 + \varepsilon_0\chi_e\right)\vec{E} = \varepsilon_0\left(1 + \chi_e\right)\vec{E} = \varepsilon_0\varepsilon_r\vec{E} = \varepsilon\vec{E},$$

where ε is the permittivity of the material and ε_r is the relative permittivity of the material. Also, the boundary conditions are now

$$\varepsilon_{\text{above}}E_{\text{above}}^{\perp} - \varepsilon_{\text{below}}E_{\text{below}}^{\perp} = \sigma_f$$

And

$$\varepsilon_{\text{above}}\frac{\partial V_{\text{above}}}{\partial n} - \varepsilon_{\text{below}}\frac{\partial V_{\text{below}}}{\partial n} = -\sigma_f$$

while we still maintain

$$V_{\text{above}} = V_{\text{below}}.$$

5.1.11 Energy in a dielectric system

Given electric field $\vec{E}$ and electric displacement $\vec{D}$, the energy in a dielectric system is

$$W = \frac{\varepsilon_0}{2}\int\varepsilon_r E^2 d\tau = \frac{1}{2}\int\vec{D}\cdot\vec{E}d\tau.$$

5.2 Problems and solutions

Problem 5.1. Given $\vec{p}_1$ and $\vec{p}_2$ below, find where to place point charge q such that there is no net torque on $\vec{p}_2$. Assume the center of $\vec{p}_1$ is the origin and express your answer in spherical coordinates.

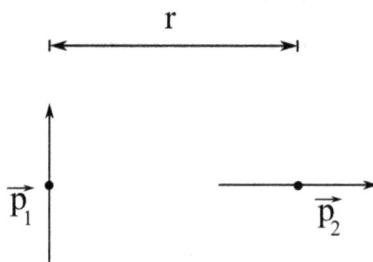

Solution The field at $\vec{p}_2$ is given by

$$\vec{E}_{\text{dip}} = \frac{p_1}{4\pi\varepsilon_0 r^3}\hat{\theta} = -\frac{p_1}{4\pi\varepsilon_0 r^3}\hat{z}.$$

Since the field is straight down, we must place the point charge q below it to cancel the field (thus resulting in zero torque on $\vec{p}_2$). We must place q at a distance d from the dipole so that

$$E_{\text{dip}} + E_q = -\frac{p_1}{4\pi\varepsilon_0 r^3} + \frac{q}{4\pi\varepsilon_0 d^2} = 0.$$

Solving for d^2 yields

$$d^2 = \frac{qr^3}{p_1}.$$

Therefore we have

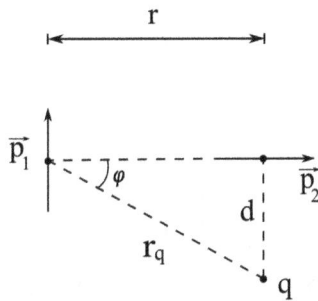

with

$$r_q^2 = \frac{qr^3}{p_1} + r^2 = r^2\left(\frac{qr}{p_1} + 1\right)$$

so

$$r_q = r\sqrt{\frac{qr}{p_1} + 1}.$$

From

$$\cos\varphi = \frac{r}{r_q}$$

we have

$$\varphi = \cos^{-1}\left(\frac{1}{\sqrt{\dfrac{qr}{p_1} + 1}}\right).$$

So the spherical coordinates of q are

$$(r, \theta) = \left(r\sqrt{\frac{qr}{p_1} + 1}, \ \frac{\pi}{2} + \cos^{-1}\left(\frac{1}{\sqrt{\frac{qr}{p_1} + 1}} \right) \right).$$

Problem 5.2. Consider a neutral atom, with polarizability α, located z above a disk of radius R carrying surface charge σ. Find the force of attraction between the atom and the plate.

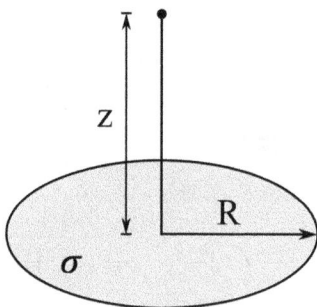

Solution The field at z is given by

$$\vec{E} = \frac{1}{4\pi\varepsilon_0} \int \frac{\sigma}{r^2} \hat{r} \ da = \frac{\sigma}{4\pi\varepsilon_0} \int_0^R \frac{2\pi r z}{\left(z^2 + r^2\right)^{3/2}} \hat{z} \ dr$$

$$\vec{E} = \frac{z\sigma}{2\varepsilon_0}\left(\frac{1}{z} - \frac{1}{\sqrt{R^2 + z^2}} \right)\hat{z}.$$

This induces a dipole

$$\vec{p} = \alpha\vec{E} = \frac{z\sigma\alpha}{2\varepsilon_0}\left(\frac{1}{z} - \frac{1}{\sqrt{R^2 + z^2}} \right)\hat{z}.$$

The electric field due to the dipole is given by

$$\vec{E}_{\text{dip}} = \frac{p}{4\pi\varepsilon_0 r^3}\left(2\cos\theta \ \hat{r} + \sin\theta \ \hat{\theta} \right)$$

$$= \frac{z\sigma\alpha}{8\pi\varepsilon_0^2 r^3}\left(\frac{1}{z} - \frac{1}{\sqrt{R^2 + z^2}} \right)\left(2\cos\theta \ \hat{r} + \sin\theta \ \hat{\theta} \right).$$

The force on a piece of charge dq is given by

$$d\vec{F} = \vec{E}dq$$

so

$$\vec{F} = \int d\vec{F} = \int \vec{E}dq = \int \vec{E}\sigma\, dA = \int \vec{E}\sigma\, 2\pi\ell\, d\ell.$$

Consider the following

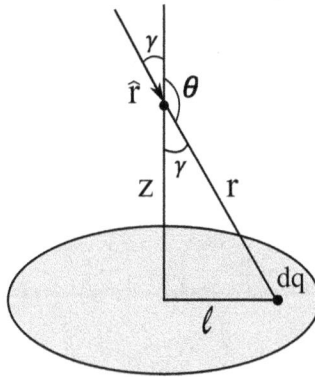

with side view

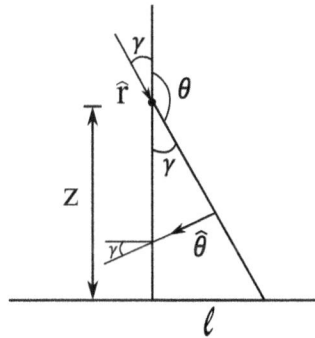

In our expression fix alignment for $\vec{E}$, we have the term $(2\cos\theta\, \hat{r} + \sin\theta\, \hat{\theta})$. Due to symmetry, we only have a $\hat{z}$-component of the force. Therefore,

$$\hat{r} \rightarrow -\cos\gamma\, \hat{z} = -\frac{z}{r}\hat{z}$$

and

$$\hat{\theta} = -\sin\gamma\,\hat{z} = -\frac{\ell}{r}\hat{z},$$

where

$$r = \sqrt{z^2 + \ell^2}.$$

Also, we have $\theta = \pi - \gamma$. So

$$\cos\theta = \cos(\pi - \gamma) = \cos(\pi)\cos(-\gamma) + \sin(\pi)\sin(-\gamma) = -\cos\gamma = -\frac{z}{r}$$

and

$$\sin\theta = \sin(\pi - \gamma) = \sin(\pi)\cos(-\gamma) + \cos(\pi)\sin(-\gamma) = -\sin(-\gamma) = \sin\gamma = \frac{\ell}{r}.$$

Therefore, we have

$$2\cos\theta\,\hat{r} + \sin\theta\,\hat{\theta} = 2\left(-\frac{z}{r}\right)\left(-\frac{z}{r}\right)\hat{z} + \left(\frac{\ell}{r}\right)\left(-\frac{\ell}{r}\right)\hat{z} = -\frac{1}{r^2}\left(-2z^2 + \ell^2\right)\hat{z}.$$

Our force becomes

$$\vec{F} = \frac{2\pi z\sigma^2\alpha}{8\pi\varepsilon_0^2}\left(\frac{1}{z} - \frac{1}{\sqrt{R^2 + z^2}}\right)\int_0^R -\frac{\left(-2z^2 + \ell^2\right)\ell}{\left(z^2 + \ell^2\right)^{5/2}}\hat{z}\,\mathrm{d}\ell$$

$$\vec{F} = \frac{z\sigma^2\alpha}{4\varepsilon_0^2}\left(\frac{1}{z} - \frac{1}{\sqrt{R^2 + z^2}}\right)\left[\frac{R^2}{\left(R^2 + z^2\right)^{3/2}}\right]\hat{z} = \frac{\sigma^2\alpha R^2\left(\sqrt{R^2 + z^2} - z\right)}{4\varepsilon_0^2\left(R^2 + z\right)^2}\hat{z}.$$

So at $z = d$, the force of attraction is

$$F = \frac{\sigma^2\alpha R^2\left(\sqrt{R^2 + d^2} - d\right)}{4\varepsilon_0^2\left(R^2 + d\right)^2}.$$

We can verify this using

$$\vec{F} = (\vec{p}\cdot\nabla)\vec{E}$$

Note

$$\vec{p}\cdot\nabla = \frac{\sigma\alpha}{2\varepsilon_0}\left(1 - \frac{z}{\sqrt{R^2 + z^2}}\right)\frac{\partial}{\partial z}$$

and

$$\vec{F} = (\vec{p} \cdot \nabla)\vec{E} = \frac{\sigma\alpha}{2\varepsilon_0}\left(1 - \frac{z}{\sqrt{R^2 + z^2}}\right)\frac{\partial}{\partial z}\left[\frac{\sigma}{2\varepsilon_0}\left(1 - \frac{z}{\sqrt{R^2 + z^2}}\right)\hat{z}\right]$$

$$\vec{F} = -\frac{z\sigma^2\alpha}{4\varepsilon_0^2}\left(\frac{1}{z} - \frac{1}{\sqrt{R^2 + z^2}}\right)\left[\frac{R^2}{\left(R^2 + z^2\right)^{3/2}}\right]\hat{z} = -\frac{\sigma^2\alpha R^2\left(\sqrt{R^2 + z^2} - z\right)}{4\varepsilon_0^2\left(R^2 + z\right)^2}\hat{z},$$

which is equal in magnitude and opposite in direction to what was found above. Why is this? In the first method, we calculated the force on the plate *from* the atom. So a positive force is 'attractive'. In the second method, we are finding the force on the atom from the plate. A negative force at the atom 'attracts' it to the plate.

Problem 5.3. Consider $\vec{p}_1$ and $\vec{p}_2$ below. Find the force of $\vec{p}_2$ on $\vec{p}_1$ and verify using the energy of the configuration.

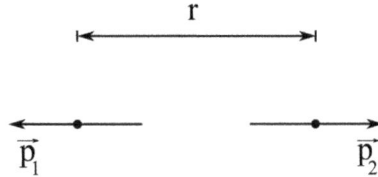

Solution The electric field due to a dipole is

$$\vec{E}_{\text{dip}}(r, \theta) = \frac{p}{4\pi\varepsilon_0 r^3}\left(2\cos\theta\,\hat{r} + \sin\theta\,\hat{\theta}\right).$$

Here, $\theta = \pi$, so the field at $\vec{p}_1$ due to $\vec{p}_2$ is

$$\vec{E}_{\text{dip}} = \frac{-2p_2}{4\pi\varepsilon_0 r^3}\hat{r} = \frac{-2p_2}{4\pi\varepsilon_0 y^3}\hat{y}.$$

The force is given by

$$\vec{F} = \left(\vec{p} \cdot \nabla\right)\vec{E}$$

where $\vec{p} = \vec{p}_1 = -p_1\hat{y}$. So

$$\vec{p} \cdot \nabla = -p_1\frac{\partial}{\partial y}.$$

Therefore,

$$\vec{F} = (\vec{p} \cdot \nabla)\vec{E} = -p_1 \frac{\partial}{\partial y}\left(\frac{-2p_2}{4\pi\varepsilon_0 y^3}\hat{y}\right) = \frac{2p_1 p_2}{4\pi\varepsilon_0} \frac{\partial}{\partial y}\left(y^{-3}\right)\hat{y}$$

$$\vec{F} = -\frac{3p_1 p_2}{2\pi\varepsilon_0 y^4}\hat{y}.$$

The energy stored in this configuration is given by

$$U = -\vec{p}_1 \cdot \vec{E} = -\left[(-p_1\hat{y}) \cdot \left(\frac{-2p_2}{4\pi\varepsilon_0 y^3}\hat{y}\right)\right] = -(-p_1)\left(\frac{-2p_2}{4\pi\varepsilon_0 y^3}\right) = -\frac{2p_1 p_2}{4\pi\varepsilon_0 y^3}.$$

From this, the force is given by

$$\vec{F} = -\nabla U = -\nabla\left(\frac{-2p_1 p_2}{4\pi\varepsilon_0 y^3}\right) = \frac{2p_1 p_2}{4\pi\varepsilon_0} \frac{\partial}{\partial y}\left(y^{-3}\right)\hat{y}$$

$$\vec{F} = -\frac{3p_1 p_2}{2\pi\varepsilon_0 y^4}\hat{y}$$

as expected.

Problem 5.4. Consider the two dipoles depicted below. Find the angle γ that maximizes and minimizes the magnitude of the torque on $\vec{p}_2$ due to $\vec{p}_1$.

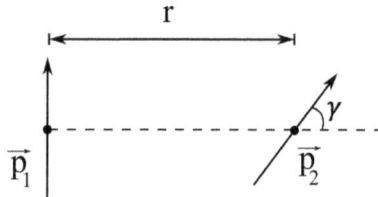

Solution The electric field due to a dipole is

$$\vec{E}_{\text{dip}}(r, \theta) = \frac{p}{4\pi\varepsilon_0 r^3}\left(2\cos\theta\,\hat{r} + \sin\theta\,\hat{\theta}\right).$$

Here, $\theta = \frac{\pi}{2}$, so the field at $\vec{p}_2$ due to $\vec{p}_1$ is

$$\vec{E}_{\text{dip}} = \frac{p_1}{4\pi\varepsilon_0 r^3}\hat{\theta},$$

which points down. The torque is given by

$$\vec{N} = \vec{p}_2 \times \vec{E}_{\text{dip}}.$$

We can express $\vec{E}_{\text{dip}}$ and $\vec{p}_2$ by

$$\vec{E}_{\text{dip}} = -\frac{p_1}{4\pi\varepsilon_0 y^3}\hat{z}$$

and

$$\vec{p}_2 = p_2 \cos\gamma\,\hat{y} + p_2 \sin\gamma\,\hat{z}.$$

So

$$\vec{N} = \vec{p}_2 \times \vec{E}_{\text{dip}} = \begin{vmatrix} \hat{x} & \hat{y} & \hat{z} \\ 0 & p_2\cos\gamma & p_2\sin\gamma \\ 0 & 0 & -\dfrac{p_1}{4\pi\varepsilon_0 y^3} \end{vmatrix} = \frac{-p_1 p_2 \cos\gamma}{4\pi\varepsilon_0 y^3}\hat{x}.$$

We can see that $|\vec{N}|$ is maximum when $\gamma = 0$ and $\gamma = \pi$

$$\left|\vec{N}\right| = \frac{p_1 p_2}{4\pi\varepsilon_0 y^3}$$

and minimum when $\gamma = \frac{\pi}{2}$ and $\gamma = \frac{3\pi}{2}$

$$|\vec{N}| = 0.$$

Note the effect of aligning the dipole parallel to the field, $|\vec{p} \times \vec{E}| = 0$, and aligning the dipole perpendicular to the field $|\vec{p} \times \vec{E}| = pE$.

Problem 5.5. A sphere of radius R carries polarization $\vec{P} = \frac{k}{r}\hat{r}$, where k is a constant, from $r = a$ to $r = R$. Find the electric field in all regions.

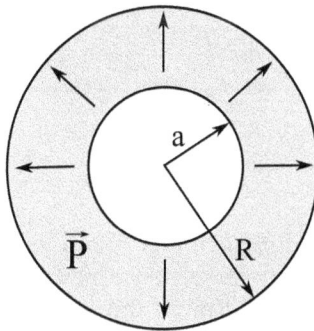

Solution The bound charges are given by

$$\sigma_b(a) = \vec{P} \cdot \hat{n} = \frac{k}{a}\hat{r} \cdot (-\hat{r}) = -\frac{k}{a}$$

$$\sigma_b(R) = \vec{P} \cdot \hat{n} = \frac{k}{R}\hat{r} \cdot \hat{r} = \frac{k}{R}$$

$$\rho_b = -\nabla \cdot \vec{P} = -\frac{1}{r^2}\frac{\partial}{\partial r}\left(r^2\frac{k}{r}\right) = -\frac{k}{r^2}.$$

When $r < a$, $q_{enc} = 0$, so $\vec{E} = 0$. When $a \leqslant r \leqslant R$, we have

$$\oint_S \vec{E} \cdot d\vec{a} = \frac{q_{enc}}{\varepsilon_0}$$

with

$$q_{enc} = -\frac{k}{a}\left(4\pi a^2\right) + 4\pi \int_a^r -\frac{k}{(r')^2}(r')^2 dr' = -4\pi kr.$$

So

$$\oint_S \vec{E} \cdot d\vec{a} = E4\pi r^2 = -\frac{4\pi kr}{\varepsilon_0}$$

and

$$\vec{E} = -\frac{k}{\varepsilon_0}\frac{1}{r}\hat{r} = -\frac{k}{\varepsilon_0}\frac{\vec{r}}{r^2}.$$

When $r > R$, we have

$$q_{enc} = -\frac{k}{a}\left(4\pi a^2\right) + 4\pi \int_a^R -\frac{k}{(r')^2}(r')^2\, dr' + \frac{k}{R}\left(4\pi R^2\right) = 0$$

so $\vec{E} = 0$.

Problem 5.6. Consider a very long cylinder of radius R hollowed out to a radius a and carrying a uniform, radial polarization $\vec{P}$ and charge density $\rho = ks$, where k is a constant. Find the electric field in all three regions ($\vec{P}$ from a to R, ρ from 0 to R).

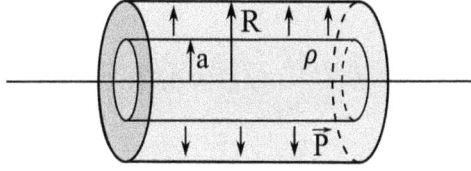

Solution For $s < a$, all we have is charge, so

$$\oint_S \vec{E} \cdot d\vec{a} = \frac{q_{enc}}{\varepsilon_0},$$

where

$$q_{enc} = 2\pi\ell \int_0^s ks's'ds' = \frac{2\pi\ell ks^3}{3}.$$

So

$$\oint_S \vec{E} \cdot d\vec{a} = E2\pi s\ell = \frac{2\pi\ell ks^3}{3\varepsilon_0}$$

and

$$\vec{E} = \frac{s^2 k}{3\varepsilon_0}\hat{s}.$$

For $a \leqslant s \leqslant R$, we have bound charge

$$\sigma_b(a) = \vec{P} \cdot \hat{n} = P\hat{s} \cdot (-\hat{s}) = -P$$

$$\rho_b = -\nabla \cdot \vec{P} = -\frac{1}{s}\frac{\partial}{\partial s}(sP) = -\frac{P}{s}.$$

So,

$$q_{enc} = -P2\pi a\ell + \frac{2\pi\ell ks^3}{3} + 2\pi\ell \int_a^s -\frac{P}{s'}s'\,ds' = 2\pi\ell\left(\frac{ks^3}{3} - Pa - Ps + Pa\right)$$

$$q_{enc} = 2\pi\ell s\left(\frac{ks^2}{3} - P\right).$$

Therefore,

$$\oint_S \vec{E} \cdot d\vec{a} = E2\pi s\ell = \frac{2\pi\ell s}{\varepsilon_0}\left(\frac{ks^2}{3} - P\right)$$

So

$$\vec{E} = \frac{1}{3\varepsilon_0}\left(ks^2 - 3P\right)\hat{s}.$$

For $s > R$, we have

$$\sigma_b(R) = \vec{P} \cdot \hat{n} = P\hat{s} \cdot \hat{s} = P.$$

So,

$$q_{enc} = -P2\pi a\ell + \frac{2\pi\ell kR^3}{3} + 2\pi\ell \int_a^R -\frac{P}{s'}s' \, ds' + P2\pi R\ell$$

$$= 2\pi\ell\left(\frac{kR^3}{3} - Pa - PR + Pa + PR\right)$$

$$q_{enc} = \frac{2\pi\ell kR^3}{3}.$$

Therefore,

$$\oint_S \vec{E} \cdot d\vec{a} = E2\pi s\ell = \frac{2\pi\ell kR^3}{3\varepsilon_0}$$

and

$$\vec{E} = \frac{kR^3}{3s\varepsilon_0}\hat{s}.$$

Problem 5.7. Consider a cylinder of radius R and length L, carrying polarization $\vec{P} = P\hat{z}$. Find the potential d from the cylinder.

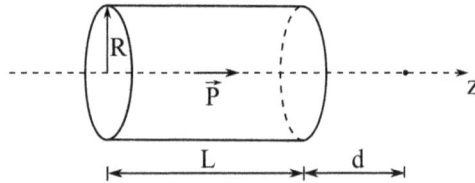

Solution The potential is

$$V(\vec{r}) = \frac{1}{4\pi\varepsilon_0} \int_v \frac{\hat{r} \cdot \vec{P}}{r^2}d\tau.$$

We can see from

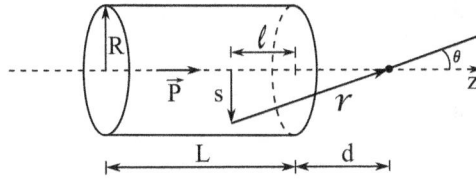

that

$$r = \sqrt{(\ell + d)^2 + s^2}.$$

If $z = 0$ is the left side of the cylinder, then ℓ goes from L to 0. So

$$r = \sqrt{(L - z + d)^2 + s^2}.$$

Also,

$$\hat{r} = \cos\theta \, \hat{z} = \frac{\ell + d}{r}\hat{z} = \frac{L - z + d}{r}\hat{z}.$$

Since $d\tau = s \, ds \, d\phi \, dz$, we have

$$V = \frac{1}{4\pi\varepsilon_0} \int_0^L \int_0^R \int_0^{2\pi} \frac{P(L - z + d)s}{\left[(L - z + d)^2 + s^2\right]^{3/2}} d\phi \, ds \, dz.$$

Using $u = L - z + d$, we have $du = -dz$. Evaluating u at the endpoints yields

$$u(z = 0) = L + d$$

And

$$u(z = L) = d.$$

So

$$V = \frac{P}{2\varepsilon_0} \int_d^{L+d} \int_0^R \frac{us}{\left(u^2 + s^2\right)^{3/2}} ds \, du = \frac{P}{2\varepsilon_0} \int_d^{L+d} \left(1 - \frac{u}{\sqrt{R^2 + u^2}}\right) du.$$

Using $x = R^2 + u^2$, we have $dx = 2u\, du$ and

$$V = \frac{P}{2\varepsilon_0}\left(u\big|_d^{L+d} - \int_{R^2+d^2}^{R^2+(L+d)^2} \frac{1}{2}x^{-\frac{1}{2}}dx\right)$$

$$= \frac{P}{2\varepsilon_0}\left[L + d - d - \sqrt{R^2 + (L + d)^2} + \sqrt{R^2 + d^2}\right]$$

$$V = \frac{P}{2\varepsilon_0}\left[L - \sqrt{R^2 + (L + d)^2} + \sqrt{R^2 + d^2}\right].$$

Problem 5.8. Consider a sphere of radius R carrying polarization $\vec{P}(\vec{r}) = kr^n\hat{r}$ where n is an integer and k is a constant. Find the charge density required to cancel the polarization.

Solution The bound charge is given by

$$\rho_b = -\nabla \cdot \vec{P} = -\frac{1}{r^2}\frac{\partial}{\partial r}\left(r^2 kr^n\right)$$

$$= -\frac{k}{r^2}\frac{\partial}{\partial r}\left(r^{n+2}\right) = -\frac{k(n+2)}{r^2}r^{n+1} = -k(n+2)r^{n-1}.$$

Therefore, a charge density

$$\rho = k(n+2)r^{n-1}$$

will cancel the bound charge produced by $\vec{P}(\vec{r}) = kr^n\hat{r}$.

Problem 5.9. A spherical shell of radius R with surface charge density σ is surrounded up to radius a by an LIH dielectric material of susceptibility χ_e. Find the electric displacement and the electric field.

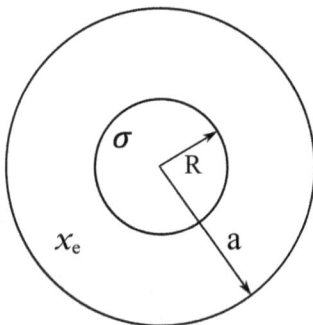

Solution Gauss's law for electric displacement is given by

$$\oint_S \vec{D} \cdot d\vec{a} = q_{f_{enc}}.$$

For $r < R$, we have

$$q_{f_{enc}} = 0.$$

Therefore,

$$D = 0.$$

For $r > R$, the left-hand side of Gauss's law is given by

$$\oint_S \vec{D} \cdot d\vec{a} = \oint_S D\, da = D \oint_S da = D4\pi r^2.$$

Now, the enclosed free charge is

$$q_{f_{enc}} = \sigma 4\pi R^2.$$

So

$$D4\pi r^2 = \sigma 4\pi R^2$$

and

$$D = \frac{\sigma R^2}{r^2} \rightarrow \vec{D} = \frac{\sigma R^2}{r^2}\hat{r}.$$

Let us consider the electric field. For $r < R$, we have

$$\vec{E} = 0$$

Since

$$\vec{D} = \varepsilon\vec{E},$$

for $R < r < a$, the electric displacement is

$$D = \frac{\sigma R^2}{r^2},$$

so the electric field is

$$\vec{E} = \frac{\vec{D}}{\varepsilon} = \frac{\vec{D}}{\varepsilon_0 \varepsilon_r} = \frac{\vec{D}}{\varepsilon_0\left(1 + \chi_e\right)} = \frac{\sigma R^2}{\varepsilon_0\left(1 + \chi_e\right)r^2}\hat{r}.$$

Finally, for $r > a$, the displacement is

$$D = \frac{\sigma R^2}{r^2}$$

so

$$\vec{E} = \frac{\vec{D}}{\varepsilon_0} = \frac{\sigma R^2}{\varepsilon_0 r^2}\hat{r}.$$

Problem 5.10. For the previous problem, calculate the electric potential everywhere, relative to infinity.

Solution The electric field in the three regions is given by

$$E = \begin{cases} 0 & r < R \\ \dfrac{\sigma R^2}{\varepsilon_0(1 + \chi_e)r^2} & R < r < a \\ \dfrac{\sigma R^2}{\varepsilon_0 r^2} & r > a \end{cases}.$$

For $r > a$,

$$V = -\int_{\infty}^{r} \vec{E} \cdot d\vec{\ell} = -\int_{\infty}^{r} \frac{\sigma R^2}{\varepsilon_0 r'^2}dr' = \frac{\sigma R^2}{\varepsilon_0 r'}\Big|_{\infty}^{r} = \frac{\sigma R^2}{\varepsilon_0 r}.$$

For $R < r < a$

$$V = -\int \vec{E} \cdot d\vec{\ell} = -\int_{\infty}^{a} \frac{\sigma R^2}{\varepsilon_0 r^2}dr - \int_{a}^{r} \frac{\sigma R^2}{\varepsilon_0\left(1 + \chi_e\right)r'^2}dr'$$

$$= \frac{\sigma R^2}{\varepsilon_0 a} + \frac{\sigma R^2}{\varepsilon_0(1 + \chi_e)}\left(\frac{1}{r} - \frac{1}{a}\right).$$

For $r < R$

$$V = -\int_{\infty}^{a} \frac{\sigma R^2}{\varepsilon_0 r^2}dr - \int_{a}^{R} \frac{\sigma R^2}{\varepsilon_0\left(1 + \chi_e\right)r^2}dr - \int_{R}^{0} 0\,dr$$

$$= \frac{\sigma R^2}{\varepsilon_0 a} + \frac{\sigma R^2}{\varepsilon_0\left(1 + \chi_e\right)}\left(\frac{1}{R} - \frac{1}{a}\right) = \text{const.}$$

Problem 5.11. A long cylinder of radius a carries a charge density that is proportional to the distance from the axis, $\rho = ks$, k constant. The cylinder is surrounded by rubber insulation out to a radius R. Find the electric displacement.

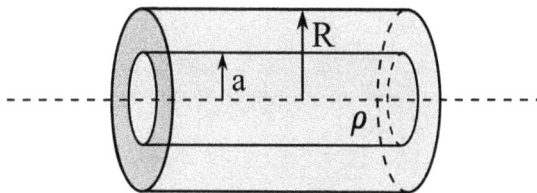

Solution For $s > a$, we have

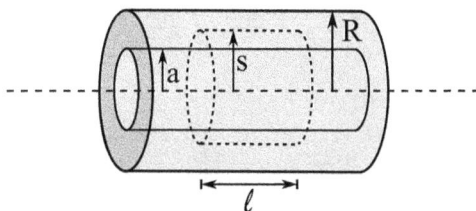

Gauss's law states

$$\oint_S \vec{D} \cdot d\vec{a} = q_{f_{enc}},$$

where the left-hand side is given by

$$\oint_S \vec{D} \cdot d\vec{a} = \oint_S D \, da = D \oint_S da = D2\pi s\ell$$

and the enclosed free charge is

$$q_{f_{enc}} = \int_{\mathcal{V}} \rho \, d\tau = k \int_0^a s^2 ds \int_0^{2\pi} d\phi \int_0^\ell dz = \frac{2\pi k a^3 \ell}{3}.$$

So

$$D2\pi s\ell = \frac{2\pi k a^3 \ell}{3}$$

And

$$D = \frac{ka^3}{3s} \rightarrow \vec{D} = \frac{ka^3}{3s}\hat{s}.$$

From this, $\vec{E} = \frac{\vec{D}}{\varepsilon_0}$ for $s > R$; if we knew $\vec{P}$, we could find the electric field as well.

For $s < a$,

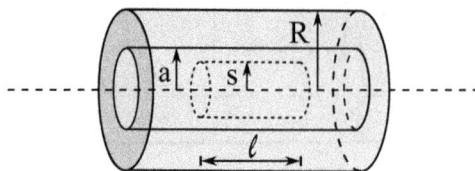

Now, the enclosed free charge is

$$q_{f_{enc}} = \int_V \rho \, d\tau = k \int_0^s s' s' ds' \int_0^{2\pi} d\phi \int_0^\ell dz = \frac{2\pi k s^3 \ell}{3}.$$

Therefore,

$$D 2\pi s\ell = \frac{2\pi k s^3 \ell}{3}$$

and

$$D = \frac{k s^2}{3} \rightarrow \vec{D} = \frac{k s^2}{3} \hat{s}.$$

The electric field can be easily obtained,

$$\vec{E} = \frac{\vec{D}}{\varepsilon_0} = \frac{k s^2}{3\varepsilon_0} \hat{s}.$$

Problem 5.12. A sphere of radius R carries a polarization $\vec{P} = k r^2 \hat{r}$, k constant:
(a) Calculate the bound charges σ_b and ρ_b.
(b) Find the electric field inside and outside the sphere by using the bound charges and Gauss's law for $\vec{E}$.
(c) Calculate the field by using Gauss's law for $\vec{D}$.

Solutions
(a) Calculate the bound charges σ_b and ρ_b.
  The bound charge density is given by

$$\rho_b = -\nabla \cdot \vec{P} = -\left[\frac{1}{r^2} \frac{\partial}{\partial r} \left(r^2 k r^2 \right) \right] = -4kr$$

and the surface charge density is given by

$$\sigma_b = \vec{P} \cdot \hat{n} = k r^2 \hat{r} \cdot \hat{n}|_{r=R} = k R^2 = \text{const.}$$

(b) Find the electric field inside and outside the sphere by using the bound charges and Gauss's law for $\vec{E}$.

What is the total bound charge (what do you expect it to be)?

$$q_b = \sigma_b(\text{area}) + \int_V \rho_b \, d\tau = kR^2 4\pi R^2 + \int_0^R \left(-4kr\right)r^2 \, dr \int_0^{2\pi} d\phi \int_0^\pi \sin\theta \, d\theta$$

$$= 4\pi kR^4 - 4\pi 4k \frac{r^4}{4}\Big|_0^R = 4\pi kR^4 - 4\pi kR^4 = 0.$$

For $r < R$, we have Gauss's law

$$\oint_S \vec{E} \cdot d\vec{a} = \frac{q_{\text{enc}}}{\varepsilon_0}$$

with the left-hand side given by

$$\oint_S \vec{E} \cdot d\vec{a} = E4\pi r^2$$

and the enclosed charge

$$q_{\text{enc}} = \int_V \rho_b \, d\tau = \int_0^r \left(-4kr'\right)r'^2 dr' \int_0^{2\pi} d\phi \int_0^\pi \sin\theta \, d\theta$$

$$= -4\pi 4k \frac{r'^4}{4}\Big|_0^r = -4\pi kr^4.$$

Therefore,

$$E4\pi r^2 = \frac{-4\pi kr^4}{\varepsilon_0}$$

with

$$E = \frac{kr^2}{\varepsilon_0} \rightarrow \vec{E} = \frac{kr^2}{\varepsilon_0}\hat{r}.$$

For $r > R$, we found $q_{\text{enc}} = 0$ so $\vec{E} = 0$.

(c) Calculate the field by using Gauss's law for $\vec{D}$.

To find the electric displacement, we consider Gauss's law for dielectrics

$$\oint_S \vec{D} \cdot d\vec{a} = q_{f_{\text{enc}}}.$$

However, we have no free charge, $q_{f_{enc}} = 0$, and $\vec{D} = 0$. Therefore, from

$$\vec{D} = \varepsilon_0 \vec{E} + \vec{P}$$

we have

$$\vec{E} = -\frac{\vec{P}}{\varepsilon_0}.$$

So

$$\vec{E} = -\frac{kr^2 \hat{r}}{\varepsilon_0} \ (r < R)$$

and

$$\vec{E} = 0 \ (r > R)$$

as expected.

Problem 5.13. A spherical conductor of radius R carries a surface charge density σ. The sphere is surrounded by alinear homogenous dielectric of susceptibility χ_e. Calculate the energy of this configuration.

Solution The energy is given by

$$W = \frac{\varepsilon_0}{2} \int \varepsilon_r E^2 d\tau = \frac{1}{2} \int \vec{D} \cdot \vec{E} d\tau.$$

It is very easy to obtain $\vec{D}$ and $\vec{E}$. For $r < R$, we have Gauss's law

$$\oint_S \vec{E} \cdot d\vec{a} = \frac{q_{enc}}{\varepsilon_0}$$

with $q_{enc} = 0$ so $\vec{E} = 0$. Also,

$$\oint_S \vec{D} \cdot d\vec{a} = q_{f_{enc}} = 0$$

So

$$\vec{D} = 0.$$

For $r > R$, the total enclosed charge is simply

$$q_{f_{enc}} = \sigma 4\pi R^2$$

and

$$\oint_S \vec{D} \cdot d\vec{a} = D4\pi r^2.$$

So

$$D4\pi r^2 = \sigma 4\pi R^2$$

With

$$D = \frac{\sigma R^2}{r^2}.$$

For $R < r < a$, the polarization is given by

$$\vec{P} = \varepsilon_0 \chi_e \vec{E}$$

so the electric displacement is

$$\vec{D} = \varepsilon_0 \vec{E} + \vec{P} = \varepsilon \vec{E}.$$

Solving for the electric field, we have

$$\vec{E} = \frac{\vec{D}}{\varepsilon} = \frac{\sigma R^2}{\varepsilon r^2} = \frac{\sigma R^2}{\varepsilon_0 \left(1 + \chi_e\right) r^2}.$$

For $r > a$, the field is just

$$E = \frac{\sigma R^2}{\varepsilon_0 r^2}.$$

Therefore, our electric displacements are

$$\vec{D} = \begin{cases} 0 & r < R \\ \dfrac{\sigma R^2}{r^2} & r > R \end{cases}$$

and our electric fields are

$$\vec{E} = \begin{cases} 0 & r < R \\ \dfrac{\sigma R^2}{\varepsilon_0(1 + \chi_e)r^2} & R < r < a \\ \dfrac{\sigma R^2}{\varepsilon_0 r^2} & r > a \end{cases}.$$

Returning to the energy, we have

$$W = \frac{1}{2} \int \vec{D} \cdot \vec{E} \, d\tau = \frac{4\pi}{2} \int_R^a \frac{\sigma^2 R^4}{\varepsilon r^4} r^2 dr + \frac{4\pi}{2} \int_a^\infty \frac{\sigma^2 R^4}{\varepsilon_0 r^4} r^2 dr,$$

where we used

$$\int d\tau = 4\pi \int r^2 dr.$$

Therefore,

$$W = \frac{2\pi\sigma^2 R^4}{\varepsilon_0}\left[\frac{1}{\varepsilon_r}\left(\frac{1}{R} - \frac{1}{a}\right) + \frac{1}{a}\right] = \frac{2\pi\sigma^2 R^4}{\varepsilon_0\varepsilon_r}\left(\frac{1}{R} + \frac{\chi_e}{a}\right) = \frac{2\pi\sigma^2 R^4}{\varepsilon_0(1+\chi_e)}\left(\frac{1}{R} + \frac{\chi_e}{a}\right).$$

Problem 5.14. A sphere of radius R, made of linear homogeneous dielectric material, is brought into a uniform electric field of magnitude $\vec{E}_0$. Using the Laplace equation and Legendre polynomials, find the electric field inside the sphere.

Solution In spherical coordinates, Laplace's equation is given by

$$\frac{1}{r^2}\frac{\partial}{\partial r}\left(r^2\frac{\partial V}{\partial r}\right) + \frac{1}{r^2 \sin\theta}\frac{\partial}{\partial\theta}\left(\sin\theta\frac{\partial V}{\partial\theta}\right) + \frac{1}{r^2(\sin\theta)^2}\frac{\partial^2 V}{\partial\phi^2} = 0.$$

We have azimuthal symmetry, therefore the potential is ϕ independent, so

$$\frac{1}{r^2}\frac{\partial}{\partial r}\left(r^2\frac{\partial V}{\partial r}\right) + \frac{1}{r^2 \sin\theta}\frac{\partial}{\partial\theta}\left(\sin\theta\frac{\partial V}{\partial\theta}\right) = 0.$$

We have outlined the solutions to this in chapter 3, and found the general solution to be given by

$$V(r, \theta) = \sum_{l=0}^{\infty}\left(A_l r^l + \frac{B_l}{r^{l+1}}\right)P_l(\cos\theta).$$

Now we can look at the boundary conditions for this particular problem. We need the electric potential to satisfy:
1) $V_{in} = V_{out}$ at $r = R$.
2) $\varepsilon_{above}E^\perp_{above} - \varepsilon_{below}E^\perp_{below} = \sigma_{free}$.
3) At large distance from the sphere: $r \gg R$, the potential must be $V_{out} = -E_0 r\cos\theta$.

Since the free surface charge density is zero, and by using the relationship between the electric field and the electric potential, the second condition becomes

$$\varepsilon\frac{\partial V_{in}}{\partial r} = \varepsilon_0\frac{\partial V_{out}}{\partial r}$$

at $r = R$.

Now with clear boundary conditions and the general solution for the potential, we can write the potential inside the sphere and the potential outside the sphere. Looking at our general solution, we require $B_l = 0$ for $r < R$; otherwise $V \to \infty$ as $r \to 0$. Similarly, we require $A_l = 0$ for $r > R$; otherwise $V \to \infty$ as $r \to \infty$. For the potential outside the sphere we want to make sure we cover the third boundary condition, and this is why we will have two terms.

Inside the sphere, we have

$$V_{\text{in}}(r, \theta) = \sum_{l=0}^{\infty} A_l r^l P_l(\cos \theta)$$

and outside the sphere, we have

$$V_{\text{out}}(r, \theta) = -E_0 r \cos \theta + \sum_{l=0}^{\infty} \frac{B_l}{r^{l+1}} P_l(\cos \theta).$$

From the first boundary condition, at $r = R$,

$$V_{\text{in}} = V_{\text{out}}$$

So

$$\sum_{l=0}^{\infty} A_l R^l P_l(\cos \theta) = -E_0 R \cos \theta + \sum_{l=0}^{\infty} \frac{B_l}{R^{l+1}} P_l(\cos \theta).$$

For $l = 1$, $P_1(\cos \theta) = \cos \theta$. So

$$A_1 R \cos \theta = -E_0 R \cos \theta + \frac{B_1}{R^2} \cos \theta$$

And

$$A_1 R = -E_0 R + \frac{B_1}{R^2}.$$

For $l \neq 1$,

$$A_l R^l = \frac{B_l}{R^{l+1}}.$$

From the second boundary condition

$$\varepsilon \frac{\partial V_{\text{in}}}{\partial r} = \varepsilon_0 \frac{\partial V_{\text{out}}}{\partial r}$$

we have

$$\varepsilon_r \sum_{l=0}^{\infty} l A_l R^{l-1} P_l(\cos \theta) = -E_0 \cos \theta - \sum_{l=0}^{\infty} \frac{(l+1) B_l}{R^{l+2}} P_l(\cos \theta).$$

For $l \neq 1$,

$$\varepsilon_r l A_l R^{l-1} = -\frac{(l+1) B_l}{R^{l+2}}$$

so we must have

$$A_l = B_l = 0.$$

For $l = 1$

$$\varepsilon_r A_1 = -E_0 - \frac{2B_1}{R^3}.$$

Consider our two equations relating A_1 and B_1,

$$A_1 R = -E_0 R + \frac{B_1}{R^2}$$

and

$$\varepsilon_r A_1 = -E_0 - \frac{2B_1}{R^3}.$$

From the first,

$$A_1 = -E_0 - \frac{B_1}{R^3},$$

and substitution into the second yields

$$\varepsilon_r \left(-E_0 - \frac{B_1}{R^3} \right) = -E_0 - \frac{2B_1}{R^3} \rightarrow B_1 \left(\frac{\varepsilon_r}{R^3} + \frac{2}{R^3} \right) = E_0 \left(\varepsilon_r - 1 \right)$$

So

$$B_1 = \frac{\varepsilon_r - 1}{\varepsilon_r + 2} R^3 E_0$$

and

$$A_1 = -\frac{3E_0}{\varepsilon_r + 2}.$$

Therefore the potential is

$$V_{\text{in}}(r, \theta) = -\frac{3E_0}{\varepsilon_r + 2} r \cos \theta.$$

Noting that $z = r \cos \theta$,

$$V_{\text{in}} = -\frac{3E_0}{\varepsilon_r + 2} z.$$

The field inside the sphere is uniform and in the same direction as $\vec{E}_0$:

$$\vec{E} = \frac{3}{\varepsilon_r + 2} \vec{E}_0.$$

Bibliography

Byron F W and Fuller R W 1992 *Mathematics of Classical and Quantum Physics* (New York: Dover)

Griffiths D J 1999 *Introduction to Electrodynamics* 3rd edn (Englewood Cliffs, NJ: Prentice Hall)

Griffiths D J 2013 *Introduction to Electrodynamics* 4th edn (New York: Pearson)

Halliday D, Resnick R and Walker J 2010 *Fundamentals of Physics* 9th edn (New York: Wiley)

Halliday D, Resnick R and Walker J 2013 *Fundamentals of Physics* 10th edn (New York: Wiley)

Jackson J D 1998 *Classical Electrodynamics* 3rd edn (New York: Wiley)

Purcell E M and Morin D J 2013 *Electricity and Magnetism* 3rd edn (Cambridge: Cambridge University Press)

Rogawski J 2011 *Calculus: Early Transcendentals* 2nd edn (San Francisco, CA: Freeman)

Chapter 6

Magnetic fields in matter

Similarly to the electric field in matter and the electric dipoles, when magnetic dipoles are subjected to a magnetic field, they may align and the medium becomes magnetized. Depending on the magnetization, we define different magnetic materials: para-magnets (the magnetization $\vec{M}$ is parallel to the magnetic field $\vec{B}$), diamagnets (the magnetization $\vec{M}$ is opposite to the magnetic field $\vec{B}$), and the special class of materials, ferromagnets, which remain magnetized even after the magnetic field becomes zero.

We would like to mention here that different sources may have different names for $\vec{B}$ and $\vec{H}$. Here $\vec{B}$ is the magnetic field and $\vec{H}$ is simply the H-field. Griffiths refers to $\vec{H}$ as the auxiliary field but we have chosen the H-field to eliminate any confusion. In other books, you may find that $\vec{B}$ is the magnetic flux density, while $\vec{H}$ is the magnetic field.

6.1 Theory

6.1.1 Torque on a magnetic dipole moment

The torque on a magnetic dipole moment $\vec{m}$ in a magnetic field $\vec{B}$ is

$$\vec{N} = \vec{m} \times \vec{B}.$$

6.1.2 Force on a magnetic dipole

The force on a magnetic dipole moment $\vec{m}$ in a magnetic field $\vec{B}$ is

$$\vec{F} = \nabla(\vec{m} \cdot \vec{B}).$$

6.1.3 H-field

Given magnetic field $\vec{B}$ and magnetization $\vec{M}$, the H-field is

$$\vec{H} = \frac{1}{\mu_0}\vec{B} - \vec{M}$$

doi:10.1088/978-1-6817-4429-2ch6

with

$$\nabla \times \vec{H} = \vec{J}_{\mathrm{f}},$$

where $\vec{J}_{\mathrm{f}}$ is the free current density. From Stoke's law,

$$\oint_S \vec{H} \cdot \mathrm{d}\vec{\ell} = I_{f_{\mathrm{enc}}},$$

where $I_{f_{\mathrm{enc}}}$ is the free current.

6.1.4 Linear media

Given H-field $\vec{H}$, the magnetization is given by

$$\vec{M} = \chi_{\mathrm{m}} \vec{H},$$

where χ_{m} is the magnetic susceptibility. The magnetic field is given by

$$\vec{B} = \mu_0 \left(\vec{H} + \vec{M} \right) = \mu_0 \left(1 + \chi_{\mathrm{m}} \right) \vec{H} = \mu \vec{H},$$

where μ is the magnetic permeability of the material, and μ_0 is the permeability of the vacuum.

6.1.5 Surface bound current due to magnetization $\vec{M}$

Given magnetization $\vec{M}$ and normal vector $\hat{n}$, the surface bound current is

$$\vec{K}_{\mathrm{b}} = \vec{M} \times \hat{n}.$$

6.1.6 Volume bound current due to magnetization $\vec{M}$

Given magnetization $\vec{M}$ and normal vector $\hat{n}$, the volume bound current is

$$\vec{J}_{\mathrm{b}} = \nabla \times \vec{M}.$$

6.2 Problems and solutions

Problem 6.1. Find the force between the two magnetic dipoles below.

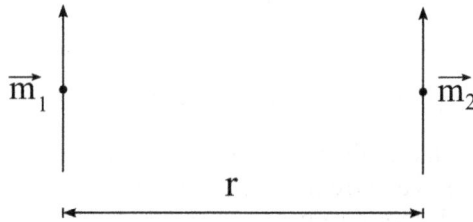

Solution The magnetic field due to dipole $\vec{m}_1$ at $\vec{m}_2$ is given by

$$\vec{B}_{\text{dip},m_1} = \frac{\mu_0 m_1}{4\pi r^3}\left(2\cos\theta\,\hat{r} + \sin\theta\,\hat{\theta}\right),$$

where $\theta = \frac{\pi}{2}$. So

$$\vec{B}_{\text{dip},m_1} = \frac{\mu_0 m_1}{4\pi r^3}\hat{\theta} = -\frac{\mu_0 m_1}{4\pi r^3}\hat{z}.$$

The force on $\vec{m}_2$ due to $\vec{B}_{\text{dip},m_1}$ is

$$\vec{F} = \nabla\left(\vec{m}_2 \cdot \vec{B}_{\text{dip},m_1}\right),$$

where

$$\vec{m}_2 \cdot \vec{B}_{\text{dip},m_1} = m_2\hat{z} \cdot \left(-\frac{\mu_0 m_1}{4\pi r^3}\hat{z}\right) = -\frac{\mu_0 m_1 m_2}{4\pi r^3}.$$

So

$$\vec{F} = \nabla\left(-\frac{\mu_0 m_1 m_2}{4\pi r^3}\right) = -\frac{\mu_0 m_1 m_2}{4\pi}\left(-3r^{-4}\right)\hat{r} = \frac{3\mu_0 m_1 m_2}{4\pi r^4}\hat{r}.$$

Problem 6.2. Find the force on a dipole located on the axis of an infinitely long cylinder of radius R, rotating at ω and carrying surface charge σ.

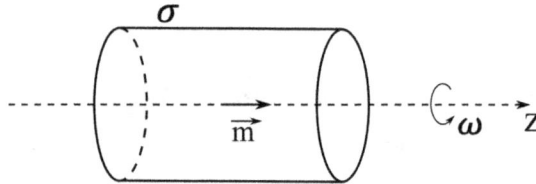

Solution We can think of the rotating cylinder as a solenoid with $nI \to K$. So

$$K = \sigma v = \sigma \omega R$$

and

$$\vec{B} = \mu_0 nI\hat{z} = \mu_0 K\hat{z},$$

which is the magnetic field inside a solenoid. The force is given by

$$\vec{F} = \nabla\left(\vec{m} \cdot \vec{B}\right) = \nabla\left(m\hat{z} \cdot \mu_0 K\hat{z}\right) = \nabla\left(\mu_0 mK\right) = 0.$$

This is an example that shows the force on a dipole in a uniform field is zero. Since we can think of a dipole as a current loop, this is equivalent to saying a current loop in a uniform field experiences zero net force.

Problem 6.3. Consider two current loops of radius R whose orientation is depicted below. Find the torque between them and the angle γ that minimizes this torque.

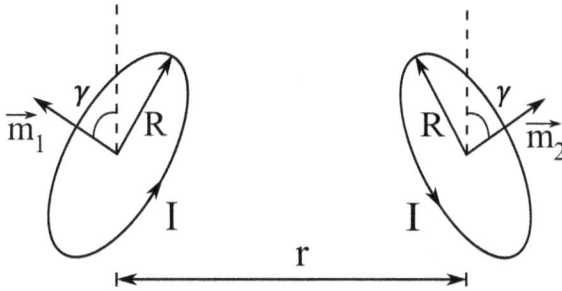

Solution Looking at this from the side, we have

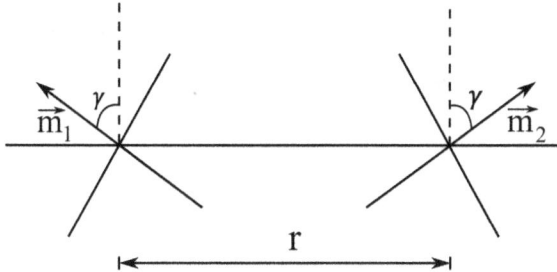

Considering $\vec{m}_1$, the field it produces is

$$\vec{B}_{\text{dip},m_1} = \frac{\mu_0 m_1}{4\pi r^3}\left(2\cos\theta\,\hat{r} + \sin\theta\,\hat{\theta}\right),$$

where $\theta = \gamma + \frac{\pi}{2}$, $r = y$, $\hat{r} = \hat{y}$, and $\hat{\theta} = -\hat{z}$. Also, we have

$$m_1 = \pi R^2 I$$

$$\cos\theta = \cos\left(\gamma + \frac{\pi}{2}\right) = -\sin\gamma$$

and

$$\sin\theta = \sin\left(\gamma + \frac{\pi}{2}\right) = \cos\gamma.$$

Using these values,

$$\vec{B}_{\text{dip},m_1} = \frac{\mu_0 \pi R^2 I}{4\pi y^3}\left(-2\sin\gamma\,\hat{y} - \cos\gamma\,\hat{z}\right) = -\frac{\mu_0 R^2 I}{4 y^3}\left(2\sin\gamma\,\hat{y} + \cos\gamma\,\hat{z}\right).$$

We have $\vec{m}_2$ given by

$$\vec{m}_2 = \pi R^2 I(\sin \gamma \, \hat{y} + \cos \gamma \, \hat{z}).$$

Now the torque is

$$\vec{N} = \vec{m}_2 \times \vec{B}_{\text{dip},m_1} = -\frac{\mu_0 R^2 I}{4y^3}\left(\pi R^2 I\right)\left[(\sin \gamma \, \hat{y} + \cos \gamma \, \hat{z}) \times (2 \sin \gamma \, \hat{y} + \cos \gamma \, \hat{z})\right].$$

Looking at just the cross product term, we have

$$(\sin \gamma \, \hat{y} + \cos \gamma \, \hat{z}) \times (2 \sin \gamma \, \hat{y} + \cos \gamma \, \hat{z}) = \begin{vmatrix} \hat{x} & \hat{y} & \hat{z} \\ 0 & \sin \gamma & \cos \gamma \\ 0 & 2\sin \gamma & \cos \gamma \end{vmatrix} = -\sin \gamma \cos \gamma \, \hat{x}.$$

Therefore,

$$\vec{N} = \vec{m}_2 \times \vec{B}_{\text{dip},m_1} = -\frac{\mu_0 \pi R^4 I^2}{4y^3}(-\sin \gamma \cos \gamma \, \hat{x}) = \frac{\mu_0 \pi R^4 I^2 \sin(2\gamma)}{8y^3}\hat{x}.$$

To find the γ that minimizes this, we must differentiate the torque with respect to γ,

$$\frac{\partial N}{\partial \gamma} = \frac{\mu_0 \pi R^4 I^2}{8y^3}\frac{\partial}{\partial \gamma}(\sin(2\gamma)) = \frac{\mu_0 \pi R^4 I^2}{4y^3}\cos(2\gamma).$$

We can find the extreme values by setting this equal to zero. Note we have

$$\frac{\partial N}{\partial \gamma} = 0$$

when

$$\cos(2\gamma) = 0,$$

which is zero when

$$2\gamma = \left(\frac{2n - 1}{2}\right)\pi$$

for a positive integer n. Solving for γ we have

$$\gamma = \left(\frac{2n - 1}{4}\right)\pi.$$

To find the minimum, we must find the second derivative of the torque. So

$$\frac{\partial^2 N}{\partial \gamma^2} = \frac{\mu_0 \pi R^4 I^2}{4y^3}\frac{\partial}{\partial \gamma}[\cos(2\gamma)] = -\frac{\mu_0 \pi R^4 I^2 \sin(2\gamma)}{2y^3},$$

Substitution of γ yields

$$\beta = -\frac{\mu_0 \pi R^4 I^2}{2y^3}\sin\left[2\left(\frac{2n - 1}{4}\right)\pi\right],$$

where a $\beta > 0$ indicates a minimum. Dropping all but the sign and the sine, we have

$$\beta = -\sin\left(n\pi - \frac{\pi}{2}\right) = -\left[\sin(n\pi)\cos\left(-\frac{\pi}{2}\right) + \cos(n\pi)\sin\left(-\frac{\pi}{2}\right)\right] = \cos(n\pi).$$

We have β is positive for $n = 2, 4, 6, \ldots$ so the torque is minimized for

$$\gamma = \left(\frac{2n-1}{4}\right)\pi,$$

when $n = 2, 4, 6, \ldots$ or for any positive integer n,

$$\gamma = \left(\frac{2(2n)-1}{4}\right)\pi = \left(n - \frac{1}{4}\right)\pi = n\pi - \frac{\pi}{4}.$$

Since each dipole moment is at an angle γ, the angle between them is

$$2\gamma = 2n\pi - \frac{\pi}{2}$$

Note the multiple of $2n\pi$ is just the addition of another complete circle. The result that minimizes the torque is a γ that causes the dipoles to be perpendicular to each other.

Problem 6.4. Consider the rotating cylindrical shell in problem 4.6, where the z-axis starts at the left side of the cylinder. Suppose we place a dipole $\vec{m} = m\hat{z}$ at a distance d from the right-hand side of the cylinder, as depicted below. Find the force on the dipole.

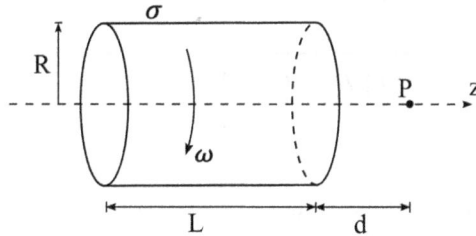

Solution From problem 4.6, the field is given by

$$\vec{B} = \frac{\mu_0 \sigma \omega R}{2}\left[\frac{d+L}{\sqrt{R^2 + (d+L)^2}} - \frac{d}{\sqrt{R^2 + d^2}}\right]\hat{z},$$

where d was the distance from our point to the right-hand side of the cylinder. We can rewrite this considering $d = z - L$. So

$$\vec{B} = \frac{\mu_0 \sigma \omega R}{2} \left[\frac{z}{\sqrt{R^2 + z^2}} - \frac{z - L}{\sqrt{R^2 + (z - L)^2}} \right] \hat{z}.$$

The force is given by

$$\vec{F} = \nabla \left(\vec{m} \cdot \vec{B} \right)$$

with

$$\vec{m} \cdot \vec{B} = \frac{\mu_0 \sigma \omega R m}{2} \left[\frac{z}{\sqrt{R^2 + z^2}} - \frac{z - L}{\sqrt{R^2 + (z - L)^2}} \right].$$

So

$$\vec{F} = \frac{\mu_0 \sigma \omega R m}{2} \nabla \left[\frac{z}{\sqrt{R^2 + z^2}} - \frac{z - L}{\sqrt{R^2 + (z - L)^2}} \right]$$

and

$$\vec{F} = \frac{\mu_0 \sigma \omega R^3 m}{2} \left[\frac{1}{\left(R^2 + z^2 \right)^{3/2}} - \frac{1}{\left[R^2 + (z - L)^2 \right]^{3/2}} \right] \hat{z}.$$

At a distance d from the right-hand side of the cylinder, $z = d + L$. Therefore

$$\vec{F} = \frac{\mu_0 \sigma \omega R^3 m}{2} \left[\frac{1}{\left[R^2 + (d + L)^2 \right]^{3/2}} - \frac{1}{\left(R^2 + d^2 \right)^{3/2}} \right] \hat{z}.$$

Problem 6.5. An infinitely long cylinder has a constant magnetization $\vec{M}$ parallel to the axis of the cylinder. Find the magnetic field due to $\vec{M}$ everywhere.

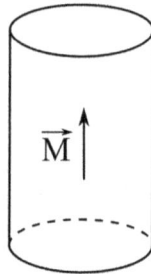

Solution The magnetization is given by

$$\vec{M} = M\hat{z}.$$

The bound volume current $\vec{J}_b$ is

$$\vec{J}_b = \nabla \times \vec{M} = 0$$

since $\vec{M}$ = constant. The bound surface current $\vec{K}_b$ is

$$\vec{K}_b = \vec{M} \times \hat{n} = M\hat{z} \times \hat{s} = M\hat{\phi}.$$

For $s < R$

$$\vec{B} = \mu_0 \vec{M} = \mu_0 M\hat{z}$$

and for $s > R$

$$\vec{B} = 0.$$

Outside, the field is zero ($\vec{B} = 0$ outside a solenoid).

Problem 6.6. A long circular cylinder of radius R has a magnetization $\vec{M} = ks\hat{\phi}$, where k is a constant, s the distance from the axis of the cylinder, and $\hat{\phi}$ the azimuthal unit vector. Find the magnetic field due to $\vec{M}$ for $s < R$ and $s > R$.

Solution Let us first find the bound volume current $\vec{J}_b$ and the bound surface current $\vec{K}_b$. The bound volume current is given by

$$\vec{J}_b = \nabla \times \vec{M} = \frac{1}{s}\frac{\partial}{\partial s}(sM_\phi)\hat{z} = \frac{1}{s}\frac{\partial}{\partial s}(ks^2)\hat{z} = \frac{1}{s}2ks\hat{z} = 2k\hat{z} = \text{const},$$

and the bound surface current is given by

$$\vec{K}_b = \vec{M} \times \hat{n} = ks(\hat{\phi} \times \hat{s})\Big|_{s=R} = -kR\hat{z}.$$

So the bound current flows up the cylinder, and returns down the surface. Let us check that the total current is zero. The total current due to the bound volume current is given by

$$I_{\text{tot},J_b} = \int \vec{J}_b \cdot d\vec{a} = \int J_b \, da = \int\limits_0^R (2k)(2\pi s \, ds) = \frac{4\pi kR^2}{2} = 2\pi kR^2$$

and the total current due to the bound surface current is

$$I_{\text{tot},K_b} = \int K_b \, d\ell = (-kR)2\pi R = -2\pi kR^2.$$

Since they are equal and opposite, the total current is zero. Now we can find the magnetic field using Ampère's law. For $s < R$,

$$\oint_S \vec{B} \cdot d\vec{\ell} = \mu_0 I_{\text{enc}}.$$

The left-hand side is given by

$$\oint_S \vec{B} \cdot d\vec{\ell} = B2\pi s$$

and enclosed current is given by

$$I_{enc} = \int_0^s J_b \, da = \int_0^s 2k 2\pi s' ds' = 2k\pi s^2.$$

Therefore,

$$\vec{B} = \mu_0 k s \hat{\phi} = \mu_0 \vec{M}.$$

For $s > R$, our enclosed current is zero, so

$$\vec{B} = 0.$$

Problem 6.7. A long cylinder of radius R carries a magnetization $\vec{M} = ks^3\hat{\phi}$, where k is a constant. Find the magnetic field due to $\vec{M}$ everywhere.

Solution Let us start by finding the bound currents. The volume bound current is given by

$$\vec{J}_b = \nabla \times \vec{M} = \frac{1}{s}\frac{\partial}{\partial s}(sM_\phi)\hat{z} = \frac{1}{s}\frac{\partial}{\partial s}(sks^3)\hat{z} = \frac{1}{s}\frac{\partial}{\partial s}(ks^4)\hat{z} = \frac{1}{s}4ks^3\hat{z} = 4ks^2\hat{z}$$

and the surface bound current is

$$\vec{K}_b = \vec{M} \times \hat{n} = ks^3(\hat{\phi} \times \hat{s})\Big|_{s=R} = -kR^3\hat{z}.$$

We can check that the total bound current is zero. From the bound volume current, we have

$$I_{tot,J_b} = \int \vec{J}_b \cdot d\vec{a} = \int_0^R \left(4ks^2\right)(2\pi s \, ds) = \frac{8\pi ks^4}{4}\Big|_0^R = 2\pi kR^4$$

and from the bound surface current, we have

$$I_{tot,K_b} = \int K_b \, d\ell = (-kR^3)2\pi R = -2\pi kR^4.$$

Therefore, the total bound current is zero, $I_b = 0$. Now we can find the field by using Ampère's law. For $s < R$,

$$\oint_S \vec{B} \cdot d\vec{\ell} = \mu_0 I_{enc}.$$

The left-hand side is

$$\oint_S \vec{B} \cdot d\vec{\ell} = B2\pi s$$

and the enclosed current is

$$I_{\text{enc}} = \int_0^s J_b \, da = \int_0^s 4ks'^2 2\pi s' \, ds' = \int_0^s 8\pi k(s')^3 ds' = 2\pi ks^4.$$

Therefore,

$$B2\pi s = \mu_0 2\pi ks^4 = \mu_0 ks^3$$

And

$$\vec{B} = \mu_0 ks^3 \hat{\phi} = \mu_0 \vec{M}.$$

For $s > R$, we have zero enclosed current. So,

$$\vec{B} = 0.$$

Problem 6.8. A sphere of radius R carries magnetization $\vec{M} = kr\hat{\phi}$. Find the magnetic field inside and outside.

Solution Since there is no free current, $\vec{H} = 0$. Inside, we have magnetization, so

$$\vec{H} = \frac{1}{\mu_0}\vec{B}_{\text{in}} - \vec{M}.$$

So $\vec{B}_{\text{in}}$ is given by

$$\vec{B}_{\text{in}} = \mu_0 \vec{M} = \mu_0 kr\hat{\phi}.$$

Outside, we have no magnetization either, so

$$\vec{B}_{\text{out}} = 0.$$

Problem 6.9. An infinitely long wire carries current I and is surrounded by material, out to radius R, with magnetization $\vec{M} = k\hat{\phi}$. Find the magnetic field for $s < R$ and $s > R$.

Solution For $s < R$, we have

$$\oint_S \vec{H} \cdot d\vec{\ell} = I_{f_{\text{enc}}}$$

with

$$\oint_S \vec{H} \cdot d\vec{\ell} = H2\pi s$$

and

$$I_{f_{\text{enc}}} = I.$$

Therefore,

$$H = \frac{I}{2\pi s}\hat{\phi}.$$

Using

$$\vec{H} = \frac{1}{\mu_0}\vec{B}_{\text{in}} - \vec{M}$$

the magnetic field inside is given by

$$\vec{B}_{\text{in}} = \mu_0\left(\vec{H} + \vec{M}\right) = \mu_0\left(\frac{I}{2\pi s} + k\right)\hat{\phi}.$$

For $s > R$, we still have $H = \frac{I}{2\pi s}\hat{\phi}$, but we do not have any magnetization. So

$$\vec{B}_{\text{out}} = \mu_0\vec{H} = \frac{\mu_0 I}{2\pi s}\hat{\phi},$$

which is exactly what we would expect from a wire carrying current I.

Problem 6.10. An infinitely long wire, of radius R carries magnetization $\vec{M} = ks^2\hat{z}$. At $s = R$, there is a surface current $\vec{K} = K_0\hat{\phi}$. Find the field for $s < R$ and $s > R$.

Solution For $s < R$, both $\vec{M}$ and $\vec{K}$ contribute to the field. We can see the contribution due to $\vec{K}$ is that of a solenoid,

$$\vec{B}_K = \mu_0 K_0\hat{z}.$$

We also have $\vec{H} = 0$, so

$$\vec{B}_M = \mu_0 ks^2\hat{z}.$$

Therefore,

$$\vec{B}_{\text{in}} = \vec{B}_K + \vec{B}_M = \mu_0\left(K_0 + ks^2\right)\hat{z}.$$

For $s > R$, we have zero magnetization, so

$$\vec{B}_M = 0.$$

Also, there is zero field outside of a solenoid, so

$$\vec{B}_K = 0.$$

Therefore,

$$\vec{B}_{\text{out}} = 0.$$

Problem 6.11. Find the H-field produced from a current density $\vec{J_f} = J_0 s\hat{z}$ in two ways.

Solution First, we will use

$$\nabla \times \vec{H} = \vec{J_f}.$$

Note, we must have $\vec{H} = H(s)\hat{\phi}$. So

$$\nabla \times \vec{H} = \frac{1}{s}\frac{\partial}{\partial s}[sH(s)]\hat{z} = J_0 s\hat{z}$$

and

$$\frac{\partial}{\partial s}[s\,H(s)] = J_0 s^2.$$

From this, we have

$$H(s) = \frac{1}{3}J_0 s^2 + C.$$

Since there is zero current at $s = 0$, we have $H(0) = 0 \rightarrow C = 0$. Therefore,

$$\vec{H} = \frac{J_0 s^2}{3}\hat{\phi}.$$

Now we will use

$$\oint_S \vec{H} \cdot d\vec{\ell} = I_{f_{enc}},$$

where

$$I_{f_{enc}} = \int J\,da = \int_0^s 2\pi s' J_0 s'ds' = \frac{2\pi J_0 s^3}{3}$$

and

$$\oint_S \vec{H} \cdot d\vec{\ell} = H2\pi s.$$

Therefore,

$$\vec{H} = \frac{J_0 s^2}{3}\hat{\phi}.$$

As expected from the first method. Note each equation we used is simply Stoke's theorem applied to the other.

Problem 6.12. This problem was inspired by a different problem presented in the Electrodynamics graduate course by Dr Charles Ebner at the Ohio State University in 2002. A sphere of radius R is uniformly polarized with a polarization $\vec{P}$. Within such a sphere, one can show that $\vec{D} = \frac{2}{3}\vec{P}$ and $\vec{E} = -\frac{\vec{P}}{3\varepsilon_0}$. By using the similarity of the equations of electrostatics and magnetostatics, find $\vec{B}$ and $\vec{H}$ within a uniformly magnetized sphere having magnetism $\vec{M}$.

Solution The equivalent equations for electrostatics and magnetostatics are the following

$$\nabla \times \vec{E} = 0$$
$$\nabla \cdot \vec{D} = 0$$
$$\vec{E} = \frac{\vec{D}}{\varepsilon_0} - \frac{\vec{P}}{\varepsilon_0}.$$

For a system with no free current,

$$\nabla \times \vec{H} = 0$$
$$\nabla \cdot \vec{B} = 0$$
$$\vec{H} = \frac{\vec{B}}{\mu_0} - \vec{M}.$$

Comparing the equations, we note that $\vec{E}$ is equivalent to $\vec{H}$, $\vec{E} \Leftrightarrow \vec{H}$; $\frac{\vec{D}}{\varepsilon_0}$ is equivalent to $\frac{\vec{B}}{\mu_0}$, $\frac{\vec{D}}{\varepsilon_0} \Leftrightarrow \frac{\vec{B}}{\mu_0}$; and $\frac{\vec{P}}{\varepsilon_0}$ is equivalent to $\vec{M}$, $\frac{\vec{P}}{\varepsilon_0} \Leftrightarrow \vec{M}$.

Starting with

$$\vec{D} = \frac{2}{3}\vec{P}$$

we can divide both sides by ε_0

$$\frac{\vec{D}}{\varepsilon_0} = \frac{2}{3\varepsilon_0}\vec{P}.$$

Since $\frac{\vec{P}}{\varepsilon_0} \Leftrightarrow \vec{M}$, we have

$$\frac{2}{3\varepsilon_0}\vec{P} \Leftrightarrow \frac{2}{3}\vec{M}.$$

Using $\frac{\vec{D}}{\varepsilon_0} \Leftrightarrow \frac{\vec{B}}{\mu_0}$, we have

$$\frac{\vec{B}}{\mu_0} = \frac{2}{3}\vec{M}.$$

Therefore,

$$\vec{B} = \frac{2\mu_0}{3}\vec{M}.$$

Now we will look at the electric field,

$$\vec{H} \Leftrightarrow \vec{E}.$$

Using $\dfrac{\vec{P}}{\varepsilon_0} \Leftrightarrow \vec{M}$,

$$\vec{E} = -\frac{\vec{P}}{3\varepsilon_0} = -\frac{\vec{M}}{3}.$$

Therefore,

$$\vec{H} = -\frac{\vec{M}}{3}.$$

Bibliography

Griffiths D J 1999 *Introduction to Electrodynamics* 3rd edn (Englewood Cliffs, NJ: Prentice Hall)

Griffiths D J 2013 *Introduction to Electrodynamics* 4th edn (New York: Pearson)

Halliday D, Resnick R and Walker J 2010 *Fundamentals of Physics* 9th edn (New York: Wiley)

Halliday D, Resnick R and Walker J 2013 *Fundamentals of Physics* 10th edn (New York: Wiley)

Jackson J D 1998 *Classical Electrodynamics* 3rd edn (New York: Wiley)

Purcell E M and Morin D J 2013 *Electricity and Magnetism* 3rd edn (Cambridge: Cambridge University Press)

www.ingramcontent.com/pod-product-compliance
Lightning Source LLC
Chambersburg PA
CBHW081532220326
41598CB00036B/6405